THE INCURABLE
LEFTIST DISEASE IN
MODERN LIBERALISM

Yuri Okunev

Translated by Jane Ann Miller

authorHOUSE®

AuthorHouse™
1663 Liberty Drive, Suite 200
Bloomington, IN 47403
www.authorhouse.com
Phone: 1-800-839-8640

First published by AuthorHouse 11/14/2008

ISBN: 978-1-4389-2708-4 (sc)

Library of Congress Control Number: 2008910014

Printed in the United States of America
Bloomington, Indiana

This book is printed on acid-free paper.

TABLE OF CONTENTS

«Something wicked this way comes»

William Shakespeare
«Macbeth»

CHAPTER I

DIAGNOSIS

There is liberalism, and there is a disease within it, called leftism. These are two different things, and they take us in opposite directions.

The author's thesis

The title of the first edition of this work – *Left-Wing Liberalism: A Senile Disorder* – was a simple variation on the title of a far more famous one written by Vladimir Ilyich Lenin: *Left-Wing Communism: An Infantile Disorder*. An older, Soviet generation no doubt remembers this book well, since they were forced to read and study it. Younger people, God bless them, are probably unacquainted with it – and with Lenin's works in general – since they are no longer forced to read anything at all.

Yet Lenin remains one of the twentieth century's most striking and sinister figures. Attitudes toward him range from abject worship to absolute rejection, but even so, no one can deny that this man who led the Bolsheviks to power in Russia in 1917 has exerted a tremendous, a monstrous, influence on twentieth-century history.

For over seventy years Lenin was idolized both within the Soviet Union and without; he was a model for millions. His works were published in hundreds of languages, in editions outnumbering even those of the Bible itself. More monuments were raised to him than perhaps to anyone else in the history of the world. Images of him once hung in every official venue in the Communist world, far outnumbering those of any other historical or mythological figure – with the possible exception of Mao. At one time, the main streets and main squares of practically every city in the Soviet Union bore Lenin's name. St. Petersburg, the former capital of the Russian Empire was renamed in his honor, as were hundreds of other towns and villages, theaters and factories, collective farms, universities, institutes, schools, libraries et cetera et cetera. His worshipers' imagination was boundless; they especially liked being able to repeat the mantra at least two or three times, as in the name of the Leningrad subway system – The Order of Lenin Leningrad Lenin Metropolitan.

However, as often happens in totalitarian societies, the collapse of the Communist state led to a radical change in attitude toward its founding fathers. Today, there are no accusations or insults too great to hurl at Lenin. Many of them are justified, for not only did he preside over the atrocities of the Red Terror, but documentary evidence has also shown the personal ferocity shown by the leader himself.

This book is not meant to be an analysis of Lenin's role in history, or of his personal human (or inhuman) qualities. So why this long discussion of him?

There is a very sound reason for taking Lenin as a starting point for a book about the grave illness currently afflicting liberal thought. The fact of the matter is that while Vladimir Ilyich Lenin was a vehement anti-liberal and caustic critic of liberalism, he and his October Revolution of 1917 mark the beginning of a new stage in the history of liberalism, a stage in which Western liberals were to be deliberately exploited by totalitarian regimes bent on undermining Western civilization and achieving world hegemony. The historical fact is simply this: before Vladimir Ilyich Lenin, totalitarian regimes saw liberalism as their chief enemy; after him, to this very day, many dictators see liberal movements in the West as potentially useful fifth columns in the camp of democracy.

Perhaps it is liberalism's misfortune rather than its fault that ever since Lenin, the powers of evil have taken such cynical and unforgivable advantage of it!

Vladimir Ilyich was characteristically cynical in his attitude toward liberals.

On the one hand, as a revolutionary, and an advocate of extreme methods in the pursuit of revolutionary goals, Lenin sincerely despised both Western and Russian liberals as "babblers who feared the violence brought on by revolution." He would soon make short work of such babblers, with the help of his Party cohort Felix Edmundovich Dzherzhinsky, the man who had found one simple, perfect, universal method of dealing with dissenters – "shoot them all." On the other hand, Lenin was perhaps one of the first to surmise how useful Western liberals might be in the continuing revolution.

As American journalist Mona Charen writes in her book *Useful Idiots*:

> Lenin is widely credited with the prediction that liberals and other weak-minded souls in the West could be relied upon to be "useful idiots" as far as the Soviet Union was concerned. Though Lenin may never have actually uttered the phrase, it was consistent with his cynical style. And, as the following chapters demonstrate, liberals managed, time after time during the Cold War, to live down to this sour prediction.

When I first read this, the application of the term "useful idiots" to liberals seemed to me both inaccurate and insulting, I didn't think the epithet even worthy of discussion, because it is common knowledge that Lenin made a practice of showering his political opponents with a wide variety of crude and gratuitous insults. Later, however, I came to understand that in this particular case neither Vladimir Lenin nor Mona Charen (who awarded him pride of place here) necessarily meant this as an insult. To the contrary, both were providing a rather accurate diagnosis of an acute illness in the body liberal.

I would like to offer my own version of that diagnosis: ***liberalism suffers from a form of moral atrophy, which manifests itself in the inability to distinguish between good and evil.***

As Mona Charen has said, and as I repeat, neither she nor Lenin was proffering a literal diagnosis. However, the expression "useful idiots" reflects perfectly the psychological disorder defined above. Lenin, the leader of the greatest attempt in human history to destroy the foundations of Western civilization and to forcibly impose a Communist ideology upon it, soon discovered how useful the leftist disorder might prove; these "useful idiots" could help the Communist regime strike the final crushing blow against bourgeois democracies already on the edge of collapse!

The forces of evil standing behind Lenin and the Bolsheviks were the very first to identify the increasingly rampant symptoms of this disorder. They did not hesitate for a moment to exploit the patient's condition.

Lenin himself had little time to put his discoveries in the field of psychiatry to practical use. Five years into the worldwide revolution he fell seriously ill; two years later, he died. But in the meantime, his ideas on the usefulness of idiots had been taken up with typical Bolshevik flourish by Iosif Vissarionovich Stalin, who styled himself both Loyal Pupil and Great Successor. Over the thirty years of his rule, Stalin used Western liberals both to solidify his own personal dictatorship and to undermine democratic regimes throughout the world.

In the long and bloody epic that was the Stalin era, we encounter some almost unbelievable passages, examples of the cynical exploitation of the illness from which liberals were suffering. But Stalin's signature work was his use of prominent liberal and world-famous writer Leon Feuchtwanger. It is worthwhile to remind readers of this case, because it reveals all the quintessential symptoms of the leftist disease, which had by that time long since ceased to be merely a "childhood illness."

The year was 1937. Stalin was busy unleashing terror throughout the nation, but meanwhile had to satisfy his political base as he solidified his power base. At the same time, he was also concerned about creating a positive image in the West. To that end, he began picking through a list of candidates, in the hope of recruiting a "fraer" from the liberal camp, someone well respected enough that the wool pulled over the fraer's eyes would cover the rest of Europe and America too.

The American equivalent for the Russian slang word "fraer" might be "mark" or "dupe"; that is, for all practical purposes, a useful idiot, a complacent and naive person whom it would be easy to recruit, to con, to deceive, but most important, one who might serve as a decoy, who would have no idea of what his true role was, but would be ever so proud of playing it.

Stalin's first choice for the lead actor in this political show was the famous French writer André Gide. An idol of the Western liberal intelligentsia, who had moreover publicly declared himself "a friend of Communism," Gide seemed perfect for the part. However, as it turned out, the choice misfired. The NKVD, which Stalin had charged with handling all delicate operations involving intellectuals, bungled the job. André Gide, who came to Moscow in 1936 and who according to Ilya Ehrenburg "admired absolutely everything he saw," went home to Paris to write a scathing criticism of the USSR entitled *Retour de l'URSS*. André Gide turned out to be neither an idiot nor useful to the cause, and at this point the whole operation seemed doomed to failure.

Stalin needed to salvage something from it, so he went to his second choice, Leon Feuchtwanger. Feuchtwanger was a world-famous writer with no party affiliations, a staunch opponent of Fascism, and a prominent liberal and humanist. Thus, Feuchtwanger was officially invited to Moscow in early 1937, and spent more than two months there.

This was at the height of the Terror. In the heart of Moscow, in the cellars underneath the Lubyanka, people were being shot by the hundreds; the bodies were hauled away in covered trucks. Any remnants of justice or rule-of-law were long gone. The country, which had broken out in a rash of forced labor camps, was in the throes of mass hysteria and terror; fits of hysterical love for the Great Leader and Teacher alternated with bizarre mass demonstrations of hatred for his supposed enemies. The country was stupefied by the sheer scale of the mindless praise for the Great Father, and crazed by the bloody reprisals wreaked on "enemies of the state"; crowds of frightened and deluded people furiously demanded that more and more "traitors" be shot. In camps on the Solovetsky Islands, in Arkhangelsk and in Norilsk, in the Urals and Siberia, in Kazakhstan, the Far East, and Kolyma, people were dying en masse. Varlaam Shalamov, a writer who by some miracle managed to survive the hell that was Kolyma, writes that

> For months on end, at morning and evening roll call, the countless lists of the condemned were read aloud. The prison band played a fanfare before and after the reading of each order of execution. The only thing that broke the darkness was the smoke from the gasoline lanterns... Hoarfrost would cover the cigarette-paper-thin orders, and the officer charged with reading them had to brush each one off with his mitten in order to make out the handwriting, and shout out the next name on the list.

Leon Feuchtwanger, the great humanist whom the government had so kindly accommodated in Moscow's best hotel, naturally knew nothing of this. This time around, the Lubyanka and its virtually illiterate and certainly amoral chief Nikolai Yezhov performed brilliantly, turning one of the bright lights of Western liberalism into just one more useful idiot. Feuchtwanger was provided with "chance encounters" on the street and in restaurants; he was allowed to attend the closed trials of such enemies of the people as Pyatakov and Radek; he was even afforded other "chance encounters" with people "critical of the Soviet government."

The final act of this remarkable show was a meeting at the Kremlin between the Leader and Teacher and his esteemed guest. Never had Feuchtwanger been so warmly greeted, so "sincerely, openly, trustingly" received. The writer was utterly captivated by the Leader – by his modesty, by a common touch combined with greatness, and a concern for the common people combined with a staunch hatred for their enemies.

Leon Feuchtwanger, struck by the chronic form of the leftist disease, did not disappoint his host, and to the very end of his life sang the song the Great Leader had written for him. After returning from the Soviet Union, he wrote *Moscow 1937*, which was first published in Amsterdam in German, with a subtitle reading "*An account of my journey for my friends*," and then reprinted the following November in a 200,000-strong (two hundred thousand!) Russian edition.

Barring some minor quibbles with life in the USSR, everything he saw there pleased him immensely.

For example, here is what he writes about the Soviet Constitution, a piece of paper Stalin could hardly be bothered to wipe his feet on, let alone more delicate parts of the anatomy:

> ....the difference between the usual constitutions of democracies and the Soviet Constitution is that while these other constitutions also declare that citizens have certain rights and freedoms, the means by which these rights and freedoms can be exercised are not pointed out; the Soviet Constitution, however, even enumerates the factors that are prerequisite for a genuine democracy.

How can one reconcile this apologia for the Soviet Constitution with the wave of lawlessness and misrule that swept over the country in 1937? How could Leon Feuchtwanger have found any "prerequisites for democracy" in the runaway cult of personality or the mass repressions that he must have seen with his own eyes? How could this writer seemingly so wise in the ways of history fail to see that the country was ruled not by law, but by the will and caprice of a malicious and merciless tyrant? Was Feuchtwanger that easily fooled? How did Stalin and those around him manage to so deftly and cynically wind such a talented, educated, and intelligent a man and writer round their collective finger? Is there any rational explanation?

Feuchtwanger concluded his account of his trip to Moscow in 1937 with a more general apologia for the Stalinist regime:

> Yes, yes, yes! How good it is, after the flaws of the West, to see a work to which one can say a heartfelt yes, yes, yes! And since I thought it would be dishonorable hide that "yes" deep in my heart, I have written this book.

How bizarre these words sound, and how unbearably painful it is to read them. They are more terrible than those barked out in the camps of Kolyma as the band played on, because these words were pronounced not by some beast in uniform, but by a leading liberal thinker! What a terrifying example of "a useful idiot," and what an ugly caricature of truthfulness and decency: a great writer and humanist justifying Stalinism, giving Stalinism a green light in the West. There has been much discussion of the inexplicable, even mysterious moral defeat Leon Feuchtwanger suffered in his intellectual bout with Stalin. Nor has there been any dearth of speculation about the motives underlying the strange behavior of this great writer: some say they were political, some say ideological, some even say nationalistic. There have been even baser suggestions, too; Stalin often paid huge honoraria to writers willing to glorify his regime, and published their works in enormous editions such as the West had never seen.

I, however, have no intention of wading into this argument; I don't plan to either defend or condemn Feuchtwanger, for the simple reason that the topic of this discussion is not Feuchtwanger at all. Our subject is the liberal movement itself, and the disease from which it suffers. Feuchtwanger was not the only apologist for Stalin's totalitarian regime: they were many; they were legion.

In his article *"Stalin's Highbrow Servants,"* historian Georgy Chernyavsky provides a detailed analysis of the activities of seven prominent Western liberals who provided invaluable service to Stalin's criminal regime: the French writer Romain Rolland, the British playwright George Bernard Shaw, the German writer Leon Feuchtwanger, the American writer Upton Sinclair, the British economists and historians Beatrice and Sidney Webb, and Hewlett Johnson, Dean of Canterbury. As Dr. Chernyavsky writes:

> And so, here we have seven public figures who lived both comfortable and enviable lives. Their longevity is testimony to the tranquility and wealth (due, of course, first and foremost to talent and hard work) in which these lives were lived. Shaw lived longest of all (he died at 94); Johnson not quite so long (92); Sinclair died at 90, Sidney Bebb at 88, his wife Beatrice at 85. Rolland lived to the age of 78, and Feuchtwanger to 74. It is unlikely that even one of them would have lived to such a venerable age had they resided in the socialist country they so praised. And besides, they loved it only from afar – the truly ridiculous thought of resettling in the USSR or applying for Soviet citizenship in order to lend a hand at building a new society never entered their minds. They simply dropped in from time to time. In reality, these people had nothing in common with [Soviet] socialism; they merely draped themselves in its exotic garb and, quite certain that nothing remotely similar would ever establish itself in their own countries, readily paid it lip service. American scholar Stephen Whitefield recalled that Feuchtwanger (then living in the US) was once asked why he hadn't emigrated to the country he seemed to regard so highly. The answer was surprisingly frank and even quite cynical: "What – do you take me for a fool?"

Stalin's highbrow servants were of course hardly fools, but the dictator considered them his useful idiots nonetheless. Here we encounter a rather common phenomenon; all too often it is not the most foolish or least educated who suffer from moral atrophy, but rather the most talented and best educated. Historians and writers alike have used up ream after ream of paper in the attempt to find a positivist-rationalist explanation of the puzzling behavior of Western intellectuals at the time: history, social conditions, et cetera – and of course, the balance of power in Europe was such that they were left with no option but to choose Stalin over Hitler. And besides, the story goes, personal circumstances often made it impossible to refuse either Stalin's favors or Stalin's money. This may be the truth, but it is not the whole truth.

In my view, the fact that Rolland, Shaw, Feuchtwanger, Sinclair, Johnson, and the Webbs (as well as French physicist Frederique Joliot-Curie, American writer Theodore Dreiser, British philosopher Isaiah Berlin and many others) joined the useful-idiots club is neither a coincidence nor a mystery. It was merely the next stage of the disease, one that that dictators can diagnose by sheer animal instinct, with no need for X-rays or MRIs.

In the post-Stalin era, the Cold War years, the illness spread. Its victims were generally people of lesser stature, although prominent writers, scholars, scientists and politicians could still be counted among the sick: the American writer Lillian Hellman, for example, or British philosopher and activist Bertrand Russell. On every stage of its drive for global dominance, up to the very day in 1991 when the Communist Party of the Soviet Union was dissolved, the billion-strong socialist camp led by the USSR actively exploited Western liberals. For more details, readers may look at Mona Charen's above-mentioned book *Useful Idiots: How Liberals Got It Wrong in the Cold War and Still Blame America First*.

As leftism metastasized, it took on ever bolder forms. The Western intelligentsia's standard justification for supporting the

Stalinist regime first came down to a protestation of ignorance: "How could we have known?" Although after 1956 it became impossible to even formally protest one's ignorance of slave labor and death camps, the spread of the illness was by then hard to contain, and the disease found newer and subtler ways to survive, i.e., "Stalin's methods are old history; we do things differently now."

It's interesting that this disorder compels its victims to constantly seek a hero, an object of morbid worship. After the fall of the Soviet Union, the leftist bacilli were in some disarray, and the mantle of the potential hero was frantically whisked from one candidate to another. Muamar Khadafi was too eccentric; Saddam Hussein was better dealt with under the table; Kim Jin Il was completely out of control; and everyone was tired of Fidel Castro. The final choice was paradoxical but ingenious. It fell on Yasser Arafat, the man who had invented virtually every weapon in the modern-day terrorist arsenal: airline hijackings, hostage-takings, suicide bombing from the air and on the ground. (Of course he had nothing in writing, no patents, but he more than once boasted that he had essentially invented hijacking and suicide bombing.)

Arafat is one of the most sinister figures of the late twentieth century, a character straight out of some phantasmagorical-historical clown show. In a fascinating article published in the *Wall Street Journal* in September 2003, Lieutenant General Ion Mihai Pacepa, a high-ranking Romanian intelligence officer who eventually defected to the West, lays out part of the story of how the Soviet KGB, at the behest of its chairman Yuri Andropov and then-dictator of Romania Nicolae Ceausescu transformed this run-of-the-mill terrorist into an internationally acclaimed "foe of American imperialism and world Zionism." Pacepa himself worked on preparing Arafat for his new job.

The KGB never did things halfway. It created a hero's biography for Arafat, even changing his place of birth from Cairo to Jerusalem; it sent him through ideological and military training at a special school on the outskirts of Moscow; it composed a propaganda brochure entitled "Our Palestine" that was later published under Arafat's own name. Andropov, the all-powerful head of the KGB, knew the uses of the disease. His obvious and cynical assumption was that usual useful idiots in the West would swallow the bait, and that they could help the KGB transform Arafat's image into that of "a courageous leader of the Palestinian people's liberation movement." Andropov's calculations proved to be 100% correct; throughout the final decade of the 20th century, monarchs, presidents, prime ministers, high-ranking Christian clerics and other influential American and European figures diligently bowed and scraped before this professional assassin. Meanwhile, the Norwegian Nobel Committee even awarded him the Nobel Peace Prize. Never before had the civilized world sunk so low. When this terrorist-laureate, whose hands were already stained with the blood of thousands, was flown from his lair in Palestine to a Paris hospital by invitation of the President of France himself, one BBC correspondent actually cried, so touched was she by "the suffering of this old, frail man."

Ekkhhh! The only explanation of this is sheer insanity brought on by the leftist disease.

Today, the epidemic continues to spread. Among the infected there number several former US presidents and vice-presidents, dozens of influential American senators and congressmen, more than half of all the political leaders in Western Europe, Canada and Israel, celebrated writers, actors and directors, countless university professors on both sides of the Atlantic, and most prominent journalists working in the mass media.

Curiously, among those struck by this acute form of the disease, we also find prominent businessmen from the global billionaires club; immeasurable wealth apparently increases the risk of infection. This applies many Hollywood stars as well, who, sitting behind the high fences of their lavish estates in sunny California, cut off from the real world, seem to have lost their moral compass.

Particularly dangerous to society are the now frequent outbreaks of leftism among judges and others in law enforcement; as their moral sense atrophies, those whose job it is to protect society against criminals are now ceasing to do so. In January 2006 the American public had the opportunity to see the extent of the decay among the "protectors" when a respectable judge from Vermont sentenced some piece of scum to just sixty days in jail for raping a six-year-old girl. Explaining his "humane" decision, the judge informed the stunned audience in the courtroom that he considered punishment senseless in general. This, ladies and gentlemen, from a judge! If this judge from Vermont were to be the only case, we might write this off to legal casuistry. Unfortunately, he is not the only case – the epidemic of leftism is spreading rapidly among both judges and attorneys. As a consequence, the legal system in the West is coming apart at the seams, and the bar of punishment for serious criminals has dropped so low that I would not be surprised to soon see a proposal to send them to some resort in Florida for rehabilitation – or perhaps to Disney World.

I neither want to nor will I name names; ever since Hippocrates, confidentiality has been one of the cornerstones of medical ethics.

On the other hand, one peculiarity of this case is that the patients themselves do not conceal their illness, but rather flaunt it, take pride in it, much like those representatives of sexual minorities who set up elaborate parades and hang banners from their balconies, putting their private lives on display for all to see.

It may be this very "openness" that has led to a massive outbreak of the disease among ordinary citizens who have no formal involvement in either politics or ideology.

Yet another peculiarity of the disease in our time is that not only individuals, but entire organizations and communities seem to be infected. Many universities in Europe and America have collectively joined the useful idiots' club. Other new members include human rights organizations. Both the former and the latter are rather successfully exploited by obscurantist and totalitarian regimes and by terrorist organizers worldwide.

Interestingly enough, in all these cases, the introduction of the bacilli leads to a standard process of deterioration. For example, some human-rights organizations that came into being as the noblest of movements created to defend people from totalitarian oppression and persecution are before our very eyes turning into accessories to the forces of evil and obscurantism. Only a few decades ago human rights activists (one of whom was Academician Andrei Sakharov, who sacrificed his career, his freedom, and perhaps his life) strove to protect dissenters from persecution by the Soviet regime. These days, their heirs strive to let thugs who have murdered innocent people (including children) go unpunished. Moreover, some of these left-liberal civil-rights organizations have become de facto legal affiliates of illegal terrorist groups.

This same process of deterioration can be observed in liberalism's attitude toward one of the modern world's central issues – the issue of socialism.

In his book *Liberalism and Economics*, Professor Alexander Movsessian contends that "liberalism, formally, pursues the same goals as does socialism, but the essential difference between the two is lies the means of achieving these goals."

Of course classic liberalism has never applied Stalinist methods (enslaving millions of workers in order to build a socialist state), or Nazi methods (annihilating whole peoples in order to build a "racially pure" socialist state).

But the malady on the left is gradually eroding the ideals of classic liberalism, erasing moral guidelines. The fine line between liberal and socialist methods of achieving a goal is being washed away by a flood of demagoguery. And now Western intellectuals are beginning to admit that, given the circumstances, Stalin's methods of building socialism were perhaps "not so bad, and perhaps the only ones possible." Then this prominent liberal or that chimes in to say that, given certain historical conditions, the end (socialism) justifies the means (totalitarianism). But we feel that we have already provided enough examples of the deterioration of liberal views on socialism in the Stalinist and immediate post-Stalinist eras.

In our day, it takes a microscope to find the difference between liberal and socialist methods of achieving the same goal. But even without the aid of magnification, we can clearly see that liberals in Europe, Canada, and a number of other countries are diligently engaged in building the sort of socialism obviously afflicted by persistent notions of social dependency and other serious complications of the leftist malady.

Once moral guidelines disappear, these infected left-liberal "humanists" start to slide down the slippery slope toward evil. Thus this disease turns honest liberals into unwitting accomplices of the forces of evil around the world, and leads to their direct complicity in the crimes committed by totalitarian regimes founded on Neo-fascist obscurantism.

Rejecting everything that is dear to healthy liberalism, the collective forces of obscurantism gleefully observe the progress of the disease of leftism within liberalism, and heartily applaud any new manifestation of it.

After reading these lines, some of my readers may decide that this author has set out to attack liberalism, liberal ideas, and liberals in general, that this author has produced yet another book blasting liberals. Nothing could be further from the truth – moreover, the author has no intention of engaging in a polemic with liberals. The object of study here is not liberalism's flaws, but the illness from which it currently suffers. Can we really assume, or even suppose, that any doctor, after discovering a serious illness in any patient, would identify the illness with the patient himself, would battle not the illness, but the patient? Of course not! It is far more likely that he will be full of sympathy for the patient and do everything in his power to prevent the disease from progressing.

There is liberalism, and there is a disease within it called leftism. These are two different things!

But in order to be convinced that this indeed the author's stance here, readers will have to wade through at least one more chapter of this work.

CHAPTER II

MEDICAL HISTORY

Syndrome: a combination of any signs and symptoms associated with any morbid process.

Stedman's Concise Medical Dictionary

Burnham depicts liberalism as more than a set of ideas: It is a syndrome that holds its adherents in a tight grip and prevents them from understanding either reality or the mindsets of those who are not liberals.

Richard Pipes

The word "liberalism" is derived from the Latin adjective "liberalis," which describes one who "is free, who yearns for freedom."

"Yearns for freedom!" How powerful, noble, and compelling that sounds!

Please allow me a sincere confession that, given the first section of this work, some readers may find rather improbable. For most of my adult life, which has spanned the latter half of the 20th century, I considered myself a liberal. And a liberal, according to any number of dictionaries, is a devotee and follower of liberalism.

"What else could I have been?" I ask, as many of my friends now look at me askance. How, in the Soviet Union, could a person be a decent human being without being at least somewhat of a liberal? It was impossible – and as proof, here is a definition from the 1979 edition of the *Soviet Encyclopedia*:

> Liberalism is a bourgeois-ideological, socio-political movement bringing together proponents of parliamentary rule, bourgeois freedoms and free capitalist enterprise.

In the Soviet Union, "bourgeois," "parliamentary" and "capitalist" were epithets hurled at anyone who diverged in the slightest from the rules of the Communist regime, at anyone democratically inclined, at anyone opposed to dictatorship. Thus anyone devoted to such values as freedom and humanism simply had to espouse liberal views. Hence from the 1960s to the 1980s, the Soviet intelligentsia breathlessly listened for any faint hint of liberal trends, and ecstatically welcomed the modest and diplomatic protests sent to the decrepit regime by Russia's patron saint of liberalism, Andrei Dmitrievich Sakharov. It eagerly supported any move toward liberalization of the power-besotted Party regime.

Like many of my contemporaries, who also lived the better part of their lives under a strict totalitarian regime, I was a natural liberal. That is, our liberalism was not a deliberate ideological choice – not some conceptual loyalty, but rather the natural reaction of a normal person to the abnormal acts of a totalitarian government. My liberalism was in essence a protest against the dogma imposed by the regime.

In justifying my liberal past, please let me say this: in the Soviet Union the word "liberal" actually meant something.

As Churchill once said, "Any 20-year-old who isn't a liberal doesn't have a heart, and any 40-year-old who isn't a conservative doesn't have a brain."

But all joking aside, there is a tendency to become more conservative with age, and if we were to track people from one generation to the next we would see a certain pattern: liberalism-to-conservatism, liberalism-to-conservatism... Occasionally, though, when the leftist disease strikes someone in old age, we see that pattern broken.

Of course, liberal thought was attractive to intellectuals in the USSR not only because it was derided in officially approved dictionaries.

Here is a traditional definition widely available on the Internet:

Liberalism: a social movement that

1) promulgates the freedom of the individual in all spheres of life, as a condition of social development;

2) fosters (in economics) free enterprise and competition;

3) fosters (in politics) the rule of law, parliamentary democracy, and the expansion of political and civil rights and liberties.

Who would raise a hand against freedom? Who would raise a hand against democracy? Who would stand against civil and political rights for all people? What civilized person would not sympathize with liberalism as defined above? Not a one!

Liberalism's positive image has been formed over several centuries, and has, in the public mind, become the foundation of a world view shared by freethinkers. For example, here is a definition from *Vladimir Dahl's 1881 Russian dictionary*:

Liberal: a political freethinker who reasons and acts freely; one who seeks greater freedom for the people and [greater freedom] of self-governance.

So let's tip our hats, let's bend a knee, let's shed tears of joy at the purity and nobility of it all. But before we do that, let us see how some other sources define liberalism.

Here are some excerpts from an early 20th century definition of liberalism, taken from Russia's famed *Brokhaus-Efron encyclopedic dictionary*:

> Liberalism is a political stance directly opposed to that of conservatism; it strives for reform and for a government and society based on individual freedoms: freedom from restrictions imposed by the church, by state despotism, police regulation, by custom...

> Political liberalism... is founded in the concept of civil liberties: the sovereignty of the people; personal privacy; freedom of thought and conscience; freedom of expression and of the press; equality. Liberalism stands for democracy and constitutional protections versus absolutism, for local self-government versus bureaucratic centralization; for civil equality versus class privilege; for public participation in the justice system...

> As for economics, liberalism posits that industry should be as free as possible from government intrusion: hence its rejection of government regulation or limitation of industrial or commercial activity; hence its demand for freedom from government intervention in trade and in relations between employers and employees...

Even in this relatively early definition of liberalism there are some discordant notes, some signs of the disease in its embryonic form.

What, for example, is meant by "restrictions imposed by the church"? Do *all* the Ten Commandments fall into the list of such restrictions, or only some of them? In general, is any liberal required to follow any of the ethical norms on which our civilization is built – if after all they are merely moral restrictions?

And what is meant by "freedom from restrictions imposed by custom"? Does such freedom also include rejection of one's native traditions, or perhaps a denial of one's heritage?

The demand for "freedom of conscience" as a basic civil right also seems very vague. Usually, freedom of conscience has been interpreted as freedom of religion. If, however, we understand conscience to be the acceptance of moral responsibility for one's actions, then the above definition seems to be quite ambiguous.

I have no intention of wrangling over these various definitions of liberalism. I would simply like to point out one simple fact, which is that if one reads them outside the rosy glow given off by the frequent repetition of the word "freedom," certain questions arise, and these definitions do not answer them clearly.

These inclarities are the very ailments that we find documented in liberalism's medical history, with more to come...

One contemporary dictionary of philosophy defines liberalism as "a socio-political theory and movement based in the concept that individual freedom is a value in and of itself."

In this we hear another dissonant note in the general symphony of praise for liberalism, or, to be blunt, see yet another symptom of the leftist disease; by this I mean the contention that individual liberty "is a value in and of itself." As applied to an individual, this phrase means literally that all the given person needs to exist is personal freedom. Freedom allows him to be himself, to refuse to adapt to anyone or anything else, to do what he wants to do and live as he wants to live, without limits or restraints.

"Wait!" exclaims the reader, surprised. "What about personal moral obligations, what about the social obligations of the individual living in a group? It may be that personal freedom as a value in and of itself is to some extent justified in politics and economics, but what about those "other spheres"? If freedom is an absolute in <u>all</u> aspects of a person's life, then why should that person be constrained by any obligations to a family, a people, or a country? For example, has he the right to refuse to serve in the military when his nation is threatened?

There are countless such questions. I do not doubt that devotees of liberalism can always find some way to answer them. And I won't dispute those answers at the moment, because in this chapter I have a different goal, which is to show that in these classic definitions of liberalism – all these sunny descriptions of numerous freedoms – one can find some very problematic sunspots. Contradictions. Birthmarks that have spread, turned morbid, or even lethal to the freedom "sufficient unto itself" on which liberalism is based. Liberty like this can lead its devotees straight into a concentration camp – something we will discuss later in greater detail.

Now that with the help of a few encyclopedias we have learned what liberalism is, we can proceed to the history of the illness from which it suffers; this in turn, requires an anamnesis, a general medical history recording the ailments liberalism has suffered in the past, which are in fact the first faint symptoms of the current complaint.

According to some historians, the term first appeared in 1812, when the word "liberals" was used to describe the authors of the Spanish Constitution. Another version holds that the word was introduced by Napoleon. However, if we are not too insistent on terminology we can attribute the birth of the concept to an earlier era.

Many historians contend that the first shoots of liberalism appeared in ancient Greek and Roman culture and philosophy (they mention Homer, Pythagoras, Protagoras, Socrates, Plato, Aristotle, Epicurus, Cicero, and others). To me this connection seems strained. One can just as easily find proto-liberal ideas in the sayings of the Old Testament prophets. So we will leave that proto-liberal antiquity out of our discussion, since the former has to do with the history of liberalism in general rather than with the history of its malady.

The particular issues that interest us first arise in the 17th and 18th centuries, in the views expressed by Descartes, Spinoza, Hobbes, Locke, and other prominent philosophers, on matters of state, religion, freedom, human rights, et cetera. While finding obvious symptoms of the disease in these early works smacks of oversimplification, one can nonetheless, in hindsight, see certain seeds beginning to sprout.

Thus we should look to the 18th century for the first faint signs of the leftist disorder, symptoms which we will here term theophobia.

The Protestant Reformation did away with many of the dogmas of medieval Catholicism, but soon replaced them with many of its own. The moment Martin Luther, the great leader of the Protestant Reformation, achieved his impressive victory over Catholic doctrine, he himself became ruthlessly intolerant of any dissent. Baruch Spinoza's philosophical revolt against both new Christian doctrines and existing Jewish ones was one of the fountainheads of classical liberalism. However, neither Spinoza's protest against religious coercion nor his critique of Biblical texts was a true manifestation of theophobia, for they lacked its aggressively atheistic component.

Clear symptoms of theophobia were to make themselves felt somewhat later, in the works of French materialists and atheists of the 18th century: in Helvetius, for example, whom historians describe as a "brilliant and witty critic of religion," and in Baron d'Holbach, who provided a "sarcastically witty critique of religion." In such "wittiness" and "sarcasm" on the topic of the eternal foundations of human ethics and morality we detect the first hint of symptoms of the leftist disorder that would come to full flower within liberalism two hundred years later. This (half) wittiness, after undergoing a long evolution that included the sophisticated mockery of religion favored by the militant atheists of both Communism and Socialism, has in our day taken shape as a morbid growth on the noble tree of classic liberalism.

Unlike the symptoms of theophobia, the symptoms of social dependency (in a nicely-wrapped socialist package) did not appear within liberalism until the turn of the century.

Classic 18th-century liberalism advanced the promising idea of private entrepreneurship unhindered by government intervention. This idea is embodied in capitalism, which despite one hundred fifty years of fierce assault by Marxists and Socialists, and numerous attempts to build something better, still remains the most effective system not only of production, but apparently the most effective social system as well. In England, the home of classic liberalism, prominent economist Adam Smith developed a theory of a market economy automatically regulated by competition. Later, also in England, David Ricardo proved that capital accumulation leads to economic growth. The Manchester school of liberal economics, whose most brilliant representative was Richard Cobden, further developed these liberal ideas by attaching the idea of free trade to the notion of freedom from government intervention. Throughout most of the 19th century the concept of free and private entrepreneurial activity dominated liberal economic theory.

As we read about the economic foundations of classical liberalism, it is hard to fathom just what has happened to them. Where have they gone? Today's liberalism asserts the very opposite: that government must intervene in matters of the economy; that social equality must be attained by taxing both producers of goods and services and, in general, anyone who works, and so on. If in the days of classical liberalism a typical liberal was a proponent of capitalism, in our time the typical leftist liberal is almost indistinguishable from a socialist.

Returning to the history of liberalism, when and how did the obsessive notion of social dependency first appear?

Professor Aleksandr Movsessian finds the beginnings of the left-turn in liberal economics in the works of 19th-century British economist and philosopher John Stuart Mill, who as one admirer put it, provided the link between classic liberalism and new liberalism. However, in our view, Mill's ideas on the social responsibility of the free individual and on the connection between social progress and such responsibility, there is no sign of leftist pathogenic bacilli: these were to appear at the beginning of the 20th century, in the works of his followers. As Movsessian writes:

> By the beginning of the 20th century, the need for government regulation in the socioeconomic sector had become obvious to a significant number of liberals both in England and on the continent. This is the era in which the doctrine of new liberalism took shape in the works of James A. Hobshouse and John Hobson in England, and of John Dewey in America.

That government regulation of the economy is necessary in sectors like the defense industry and military contracting is obvious not only to liberals, but to any person with common sense – just as is the need for government to carry out certain functions in the sphere of education, social security, medical care and the like. This obvious need is not the point.

The new-liberalist doctrine was a symptom of the leftist disorder for a different reason, in that it proposed moving the idea of socialization front and center, and at this point the difference between socialism and liberalism quickly begins to fade. Early 20th-century American economist and theoretician of new liberalism John Dewey wrote:

> Earlier liberalism regarded the separate and competing action of individuals as the means to social well-being as an end. We must reverse the perspective and see that a socialized economy is the means of free individual development as the end.

If we had no idea who wrote this, and if we were to replace "earlier liberalism" with "capitalism" (a proper equivalent), Dewey's statement could easily fit into a chapter from Stalin's *Short Course of the History of the CPSU (Bolshevik)* – "The Economic Program of the Communist Party."

New-liberal notions of a regulated form of capitalism were finally refined and defined in the 1930s in the works of the English economist John Maynard Keynes. Keynesian theory would later play a decisive role in the creation of modern European socialism.

The appearance of these morbid symptoms caused a break in liberal ranks. Among those publicly critical of new-liberal ideas were prominent 20th-century philosophers and economists like Friedrich von Hayek, Ludwig von Mises, Karl Popper, and Isaiah Berlin. They warned that the socialization of an economy would inevitably lead to totalitarianism in policy. Nonetheless, these new-liberal ideas about [the need for] socialized economies have spread rapidly, invaded politics, and now figure in virtually every part of people's lives. In social and psychological terms, these ideas have taken a variety of morbid forms: forced "leveling," social dependency, an unwillingness to engage in competition, and eventually to a rejection of any sort of active life.

Throughout the history of liberalism, one can find quite a few other latent symptoms of the ailment to come. I cannot help but feel that many of the contemporary manifestations of the leftist disease lie in some very early liberal concepts. For example, what about Utilitarianism, whose founding father was the 18th-century English philosopher Jeremy Bentham? Here is how Aleksandr Movsessian, in remarking on the enormous influence this philosopher has had on the development of liberalism, summarizes the chief postulates of utilitarian theory:

- all human actions are motivated by a desire to gain pleasure and avoid pain;

- anything that obtains the greatest good for the greatest number of people is moral.

I would rather not spend time discussing these postulates: they speak for themselves. But in them, I believe, lie the sources of many of the quite cynical manifestations of the leftist disease in our day – about which further.

Through the complex and perhaps even confusing history of liberalism one can, nonetheless, track a via dolorosa from the bright peaks of classic liberalism down to the grim, noxious abyss of modern liberalism. Here is where the seat of the leftist infection lies, an infection that leads to the atrophy of the moral sense and, eventually, to the inability to distinguish between good and evil.

Discussions of those gleaming heights are always pleasant and easy; it is always nicer to sidestep dark chasms.

Nonetheless there have always been people unwilling to avert their eyes from dark and unpleasant places. One of these was the American writer and scholar James Burnham, who in 1964 published a bestseller entitled *The Suicide of the West*. Richard Pipes, commenting on the book in his own expressively titled article, "*Dissecting Modern Liberalism*," writes:

Burnham depicts liberalism as more than a set of ideas: It is a syndrome that holds its adherents in a tight grip and prevents them from understanding either reality or the mindsets of those who are not liberals.

Given that the medical definition of a syndrome is a "combination of signs and symptoms associated with any morbid process, which together constitute the picture of the disease" we can say without the slightest exaggeration that James Burnham was one of the first to diagnose this illness on the left.

In our time, many people are aware of the illness, but are embarrassed to talk about it.

So, for example, here we have a clearly ill woman onstage at an antiwar rally, frantically waving her arms; here we have a clearly ill man screaming at TV cameras. Everyone sees it, but no one says a word; somehow everyone is embarrassed to call a spade a spade. Better to pretend that nothing is wrong, that all this is merely an ordinary difference of opinion.

Ignorance is convenient both for the patient and those who surround him; ignorance comforts and calms. But the great danger of ignorance is that it masks the impending disaster, and makes it impossible to avert.

In his book *Metapolitics*, the renowned writer Igor Yefimov, who in Soviet times had to write his philosophical essays under a pseudonym (Andrei Moskovit'), examines the confrontation between unknowing and knowing.

> The gift of reason intrinsic to each and every human will can be compared to a spotlight created to illuminate the world around both in time and space. Freedom of will cannot be fully realized except through this gift. The choice lies in where to direct that spotlight: to carefully, selectively skirt anything that is frightening, critical, difficult, repulsive or dangerous is to choose ignorance; to beam a clear and constant light around oneself, never dimming it and never moving it away from even the most fearful and painful of scenes is to choose knowledge.

The writer then continues to argue that the choice of knowledge over ignorance is a fundamentally spiritual act, the very foundation of all human progress.

So shall we make this courageous choice in the battle against the leftist disorder? This choice underlies many of the symptoms that we see; denial has become part of life for many who suffer from this illness. This is not their fault; it is their misfortune.

But for the healthy, the choice of ignorance over knowledge is not a fault. It is a sin.

How long can we pretend that liberalism is not diseased?

How long can we avert our eyes from the chronically, violently ill?

How long can we, standing noble and tolerant, polemicize with those who are clearly ill?

How long can we hide the truth behind some pitiful version of political correctness?

Is it not time to give these symptoms their proper name?

Those who want to stay in the cozy and carefree kingdom of unknowing should not read any further. However, those who are ready to choose knowing should read on.

CHAPTER III

PRIMARY SYMPTOMS OF THE ILLNESS: FORTOPHOBIA

Evil exists for us to fight against it, Not to weigh it on a water-yoke.

Joseph Brodsky

The word "fortophobia" derives from the Latin "fortis," meaning strong or powerful; "fortitudo" is physical strength. The English word "force" is also derived from the same root. Thus, "fortophobia" is a fear of force or violence, and in the context of this work, signifies an obsessive fear of applying force or violence to combat evil.

The rejection of violence is an absolutely natural feeling for a modern, cultured person who is utterly opposed to Fascism, Bolshevism, or any of the other "ism" grounded in sheer force. However, when the leftist bacilli take hold, this perfectly natural feeling becomes what we call "fortophobia," a rejection of all active resistance to evil, no matter how deep-rooted, blatant or irreversible the evil might be.

Placing limits on force and violence has always been one of the fundamental principals of classic liberalism. Professor Aleksandr Movsessian, a specialist on the history of liberalism, has written:

> Liberals hold that no one has the right to use force except in the enforcement of procedures established by law and known to society as a whole. No one has the right to use force except in the defense of the rights of the individual; only when these rights have been violated can force be applied for the public or other good.

At first glance this liberal concept seems irreproachable; one's first impulse is to endorse it unconditionally. But curious things happen under cover of these irreproachable words.

A group of Muslim fanatics and suicide bombers arrives in the United States. No one has the right to ask why these people have come; no one has the right to send them home, to limit their movements around the country, or even to track what they are doing here. After all, that would constitute illegal force, since they have not (not yet, anyway) violated anyone else's rights.

Sometime later these suicidal fanatics enroll in private flight schools. No one has any right to expel them, because that would contradict "procedures established by law and known to the society as a whole." No one has any right to even investigate why they want to learn to fly a commercial airplane, since one "cannot use force except in defense of the rights of the individual, and only if those rights have already been violated..."

Meanwhile, these fanatics are planning a massive terrorist strike. But no one has any right to track their movements or trace their phone calls; no one has any right to infiltrate their organization, because "no one has any right to use force except in defense of the rights of the individual, and only if those rights have been violated..."

On September 11, 2001, these fanatics freely board commercial airliners, murder the pilots, and fly the planes into public buildings – and 3000 Americans die in the flames of these monstrous blasts.

Currently there is much talk about the mistakes made by the various Federal agencies who were responsible for the safety and security of Americans but who were unable to prevent this Islamo-Fascist attack on the country. There are truly no words to describe this failure. The incompetent and irresponsible leadership of these agencies, a leadership that had sunk into a complacent, lazy and well-fed routine, made one mistake after another. However, the root of the problem lies not in these mistakes themselves, but in the ultra-liberal American system of dealing with criminals. The assertion that the September 11 terrorist acts were unexpected ("who could have foreseen this?") is a sheer lie. In fact, the first attempt by Islamo-Fascist groups to blow up the World Trade Center was made in 1993. No serious measures to prevent further activity by these groups (in the United States at least) was undertaken, and the government limited itself to incarcerating a few of the more odious leaders in luxury resorts (sorry, but that's the only way the author can describe the comfortable American prisons where these terrorist organizers are held). Then, over the next eight years the fuehrer of global Islamo-Fascism, conveniently tucked under the wing of the obscurantist Taliban regime, organized a number of mass murders in a number of countries, killing Americans and anyone who happened to be near them. This fuehrer made no pretense of concealing his intentions; he planned to smite the infidel in the very heart of America. All that the left-liberal leaders of America could manage to do in response was mount an absurd bombing raid on a chemical plant in Sudan that supposedly belonged to the fuehrer – for which they later had to apologize. They also staged a comic-opera rocket attack in the mountains of Afghanistan, for which they did not have to apologize, because what little dust it raised did not harm the terrorists in the least. The fuehrer and his Islamo-Fascist cells throughout the world made open fun of America's idiotic behavior.

Thus, the catastrophe that was September 11 was less a result of a sudden assault by the global forces of evil and obscurantism than a consequence brought on by the Western world's refusal to use preemptive force in the interests of self-defense.

Three thousand innocent Americans lost their lives on September 11 in the name of a monstrously false concept – the rejection of force as a means of combating evil.

The symbolic finale of this farce, the apotheosis of absurdity, came with the anecdotal report that the U.S. Immigration and Naturalization Service had issued a residence visa to the leader of this Islamo-Fascist group of terrorists. The document mailed to the dead fanatic, arrived approximately one month after he flew his hijacked airliner and its 92 passengers into one of the World Trade Center towers in New York.

If any of my readers suppose that after the sad story of September 11, 2001, the attitude of America's left-liberal organizations toward use of force against potential terrorists has changed, I congratulate them on their record-breaking naiveté! The leftist disease does not let go so easily. As the Roman historian Tacitus once put it, medicines unfortunately act much more slowly than do diseases.

In February 2006, four-and-a-half years after September 11, I happened to see a television program featuring a conversation between Bill O'Reilly and a representative of an equally well-known American organization devoted to the defense of civil liberties. (Given that he and his organization are gravely ill, medical ethics does not permit me to reveal the name of either.) When the journalist asked whether Federal security agencies had the right to conduct surveillance of a person suspected of plotting a terrorist act, this "guardian of civil liberties" said no, they did not, and added, literally, that terrorists in the US should enjoy the same rights as ordinary citizens; that only after the commission of some terrorist act can force or discrimination be used against them – with the additional proviso that the commission of this act be proved in court. Need I say more?

Liberalism is by now very ill with fortophobia, a hereditary disease passed down through all movements whose medical history has contained even a microscopic dose of the concept of nonresistance to evil.

The history of this disorder is a curious one. As we will soon see, many of the complications of the leftist disease have been brought on by contact with ideological and political movements not only far removed from classic liberalism, but often antagonistic to it. This is the case with fortophobia: liberalism caught the infection from Christianity, which itself has long since recovered from the disease.

But to confirm this diagnosis, we must first make a historical digression.

The notion of passive resistance was defined nearly two thousand years ago in Jesus Christ's Sermon on the Mount. In the Gospel of Matthew, Christ preached that

> Ye have heard that it hath been said, An eye for an eye, and a tooth for a tooth: But I say unto you, That ye resist not evil, but whosoever shall smite thee on the right cheek, turn to him the other also... You have heard that it hath been said, Thou shalt love they neighbor, and hate thine enemy. But I say unto you, Love your enemies, bless them that curse you, do good to them that hate you and pray for them which despitefully use you, and persecute you.

> (Matthew 5:38-39, 43-44)

This core teaching lay at the heart of everyday life in early Christian communities – but only until Christianity became the official religion of the Roman Empire. As long as Christianity was a faith for the wretched of the earth, the notion of non-resistance to evil did not conflict with the way they lived their lives. But when the wealthy and powerful began attending church, the desire to retain that wealth and power came into irreconcilable conflict with the concept of nonviolence.

Later, over the course of many centuries, Christ's Sermon on the Mount was consigned to deliberate oblivion, and the notion of passive resistance to evil was held sacred by only the most radical of sects. After the Reformation, a handful of Protestant groups tried to resurrect the credo of nonviolence, but with little success. By the second half of the 19th century, Christ's teachings on non-resistance had been completely discredited by the practical realities of life, and were merely the old, worn-out trappings of a faith.

The great Russian writer Lev Tolstoy, the most prominent supporter and advocate of passive resistance to evil, initially looked to Russian Orthodox Christians in search of some sign, some vestige, of a heartfelt or serious belief in this core Christian ideal. But he found nothing, neither among the aristocrats whose education and way of life resembled his own, nor among the simple folk. So as the 19th century came to a close and the 20th century began, Tolstoy undertook a grandiose attempt to reform Christianity by returning to the concept of nonviolence and passive resistance. His plans for reform failed for many reasons, but in my view the chief reason for the failure of the Tolstoyan movement was that it was based on an absolutely untenable idea. Even in those blessed times, when evil was just a ripple compared to the flood of hate that would engulf Europe some fifteen years later, even in that relative paradise, the average Christian did not believe that evil could be overcome without the use of force.

Lev Nikolaevich Tolstoy died in 1910. In the decades immediately following, the peoples of Europe would be beset by a series of catastrophes, the scale of which dwarfed the entire sum of human suffering up to that point. It was clear that non-violence, as a movement, had failed.

The following observation is also of interest, in light of our topic.

In attempting to comprehend the human catastrophe that took place during World War II, many writers and scholars have posited that Auschwitz was both the real and the symbolic grave of

European civilization, and that Europe would never again attain those cultural heights it had reached in the pre-Auschwitz era.

For example, in the words of French philosopher and literary critic Philippe Lacoue-Labarthe, "God indeed died at Auschwitz, in any case, the God of the Greco-Christian West."

Spanish journalist Sebastian Vivar Rodriguez expressed it even more succinctly: "Europe died at Auschwitz."

There are many indications that these conclusions are accurate; however, we might just as accurately say that Auschwitz also, once and for all, laid to rest the Christian ideal of nonresistance to evil. The scale of the evil perpetrated at Auschwitz is such that even talking about nonresistance seems simply indecent. Even Mahatma Gandhi, the most dedicated and least comprising advocate of this concept, the most famous non-Christian follower of Lev Tolstoy, said in 1947, "...I must confess my bankruptcy... such non-violence can have no play in the altered circumstances." The great Indian humanist did not live to see Stalin's crimes exposed, but if he had, I suspect that he would have expressed himself even more deliberately. Stalin's death camps in Kolyma differed from Hitler's Auschwitz only in that they were less mechanized and efficient – not because they were more humane.

And so the Christian ideal of nonresistance to evil died at Auschwitz.

However, just as there is a law of conservation energy, there is a law of conservation of ideas. They do not disappear, but merely change their shape and their "carrier." So where has the idea of nonresistance, supposedly destroyed at Auschwitz, gone? Where has it found a new home?

Bravo, reader! You have it right: ironically enough, the bacilla migrated from Christianity to left-liberalism, where it fostered the malignant growth we call fortophobia. Why is this ironic? Because, historically, this disease has tended to immunize liberalism against all sorts of religious doctrinarism; here, however we see a direct borrowing of one of Christianity's central tenets.

And as for Europe, we see a quite depressing picture: there, fortophobia has become an irreversible condition.

To paraphrase the famous first line of Karl Marx's *The Communist Manifesto*, "A specter is haunting Europe – the specter of non-resistance."

Europe has abolished the death penalty and banned the extradition of criminals to countries where the death penalty is still in force. In Europe, intent itself is not considered a crime. For example, someone who plans to plant a bomb in the London tube and kill hundreds of people cannot be subjected to any "force" until he actually acts upon his intention.

Not a single European leader can bring himself to take forceful action against terrorists and murderers, let alone against their patrons or their admirers. The British government's reaction to the detonation of several "live bombs " (who were British citizens, by the way) in London in 2005 was truly pathetic; the prime minister, heir to the seat occupied by the great Winston Churchill, hurried to reassure local Islamic leaders that Great Britain was not hostile to Islam (God forbid.) The response of the French government to the recent riots mounted by immigrants in Paris was even more ridiculous. It appropriated additional funds for the construction of mosques.

Europe, for all intents and purposes, refused to take join the American military effort against the Islamo-Fascist regimes in Afghanistan and Iraq. France, Germany, Russia and other European countries condemned the US for its use of force against Saddam Hussein, one of the bloodiest dictators in modern history. Britain was the only European nation to support this both justified and necessary act of force.

In October 2005, a certain European prince whose family squabbles have preoccupied tabloid readers for the better part of two decades came to the United States on an official visit. How pleasant, how useful the prince's visit might have been had he just come to show off his new wife! But no! The high-ranking guest, wishing to display his "progressive and humane" side as well, asked the US president to show tolerance for Islam – or to put it bluntly, asked him refrain from using force against Islamo-Fascist terror. Note that the prince made his plea for tolerance not to the Islamo-Fascist leaders who have murdered innocent people in a number of countries across the world, but to the leader of the only country attempting to halt the spread of terror across the planet.

Fortunately, little enough depends on this prince, who is himself a victim of the leftist disorder. But if he and those like him were to infect America's leadership, it would be a catastrophe for the entire world.

I would like to make myself as clear as possible here. If those suffering from this aggravated form of the leftist disease come to power in the United States, we will find ourselves living through many more September 11-ths, and the 2001 original will pale in comparison with them.

A secret weapon in the arsenal of evil, fortophobia masks itself in various ways, but always drapes itself in the lily-white mantle of pure and untainted humanism. The two most dangerous guises disease takes are pacifism, and clemency toward criminals – to which topics we must devote separate chapters.

CHAPTER IV

SYMPTOMS OF PROGRESSION TO A CHRONIC STATE: OBSESSION WITH PEACEMAKING

You were given the choice between war and dishonor. You chose dishonor and you will have war...

Sir Winston Churchill

An appeaser is one who feeds a crocodile –hoping will eat him last.

Sir Winston Churchill

Evil generally attacks when the camp of good is busy talking about peace.

The author's conclusion

In our time, one of the most tenacious and dangerous concepts derived from the idea of nonviolence is pacifism. Modern pacifism, which suffers from an obsession with peacemaking, is a symptom of the onset of a chronic illness.

Here we encounter a process typical to the course of this particular disease, i.e., the gradual transformation of an initially noble idea into its polar opposite. Just as a healthy cell, when affected by certain physical, chemical or genetic factors, can become malignant and in turn cause the cells around it to deteriorate, the noble idea of rejecting violence turns into an absurd rejection of any right to defend oneself against deep-seated and flagrant evil.

But what's so bad about pacifism? In essence, there is nothing bad at all, and in fact there's much good in it. Some of the best pages of the Bible are devoted to the idea of peace among nations. For centuries, prominent intellectuals, philosophers, clerics, scholars, and scientists have stood up to condemn war and violence. After the First World War, Western European and American writers created a whole series of masterpieces imbued with pacifist ideas – for example, the novels of American Ernest Hemingway and German Erich Maria Remarque.

It would be absurd to accuse pacifist writers of causing the tragedy that occurred in Europe in the mid-20th century. But facts are stubborn things.

Pacifist ideas in postwar Europe and America so weakened these peoples' ability to fight evil, so demoralized them, that they were incapable of mounting any resistance to a growing movement fed and raised on appeasement – German Nazism. The democratic governments of Western Europe and America missed that fateful moment, that point of no return, at which they still might have stopped the expansion of Nazi Germany by a simple, almost surgical military strike that would have resulted in minimal casualties. Having missed it, they tumbled into the abyss of blood and slaughter that was World War II.

In the end, the mighty forces of good won the day, but the final victory over Fascism carried off almost 100 million lives, and brought widespread devastation and ruin to Europe.

Did Europe's liberal leaders know Hitler's intentions? Yes, they did.

Was it possible to avoid the bloody consequences of his acting upon his intentions? Yes, it was. It would have required using military force against Hitler, and up until the mid-1930s no more than a few divisions would have been needed.

But pacifism led the peoples of the world down the garden path of appeasement: instead of organizing an assault to remove the Nazi regime, the "peacemakers" organized the Berlin Olympics. This outrageous insult to the Olympic ideal of peace, to common sense itself, was apparently "justified" by the hopes of pacifying the Nazi beast. The Fascist leadership assessed this cowardly act of "peacemaking" for just what it was worth – and a scant two years after the "peaceful" Olympics, the Nazis had occupied Austria, Czechoslovakia, and set about the total annihilation of the Jew people. A year later, convinced that it could act with impunity, the regime unleashed the Second World War.

Did liberals learn any lesson from their not-so-distant past experience? Absolutely not – and in our times history is repeating itself to a ridiculous degree.

Islamo-Fascism (let's call a spade a spade) has issued a monstrous challenge to the entire world (the Muslim world included, by the way!): either you will live the way we want, or we will kill you wherever and whenever we want, all of you – Jews, Americans, Russians, Britons, Spaniards, French, Italians, Germans, Hindis and other "infidels," as well as an Muslims we deem not entirely devout – Iraqis, Jordanians, Afghanis, Pakistanis, Indonesians et al. Across the vast reaches of this planet huge numbers of fanatics are being organized and trained for war; the Fascist hordes of World War II look small in comparison. This new host of evil is not grouped into regiments or divisions, but instead covers the surface of the earth like an invisible scum made up of small terrorist bands united not by a central command but by a central idea and aim, which is to overthrow the "infidel" civilization they find so hateful.

And those who are expected to change their ways and submit to this obscurantism? How have they reacted? In the very same way that liberal European leaders reacted to Hitler – they exhort, they retreat, they attempt to placate the raging obscurantists by means of concessions, by restraint and by, to use a popular little phrase from the left-liberal lexicon, political correctness.

Liberal movements within the mass media everywhere opposed and continue to oppose the forceful overthrow of the obscurantist Taliban regime in Afghanistan, against the removal of the Islamo-Fascist regime of Saddam Hussein in Iraq, against almost any forcible action at all against the terrorists. Millions of pacifists try to obstruct any political figure who speaks in favor of mounting even the smallest of military operations against terrorists.

Even in the United States, which used military force to remove the totalitarian regimes in Afghanistan and Iraq – regimes supported by terror, that spawned terror all over the world – even here an enormous wave of pacifism rose up, threatening to wipe away the modest gains that, in spite of the pseudo-peacemakers, have been made in the war on terror.

In 2003, some time after the US incursion into Iraq, I happened to spend an evening in a Brooklyn restaurant where a friend of mine was celebrating his birthday... Seated next to me was a man of about fifty whom I'd occasionally seen with the other friends there, but our acquaintance was quite superficial. The talk, as usual, turned to politics – something I greatly dislike when there is vodka on the table. My attempts to steer the conversation toward a different topic met with no success, and I had to listen to my neighbor's pronouncements about how shameful it was for America to have launched this war on Iraq. I tried to turn it into a joke, saying that Saddam Hussein would agree with him completely, but he refused to take up this lighter vein. Aggressively proclaiming his pacifist views, he began using increasingly strong expressions in condemnation of the US policy in Iraq, trying to provoke an argument; apparently, he intuitively took me for an opponent.

I managed to get through all his pronouncements in defense of "the victim of American aggression" (that is, Saddam Hussein's obscurantist regime) but when he, apparently wanted to draw the rest of the table into the argument, loudly and proudly announced that he was "against all war," the Rubicon was crossed.

I took a long, intent look at my tablemate – for the first time, here was someone in the advanced stages of the leftist disease, sitting at my very elbow. Pacifist babble like "I am against any and all war!" generally tends to push me over the edge. And that's what happened that evening.

- If you're opposed to all war, then you would be opposed to a war against Hitler? Am I right?
- Hitler and his policies are products of previous wars...
- I'm not asking about products... Just answer my question. Are you for or against war with Hitler?
- I'm against the Second World War...
- I wasn't asking about the Second World War. Please answer a simple question. Are you for fighting Hitler or against fighting Hitler?
- ....
- Your silence tells me that you're against, which means that you are his accomplice, you've aided and abetted his crimes. And that explains perfectly why you're so concerned with the fate of a Fascist like Saddam Hussein.
- My opinions don't give you the right to...
- Yes, they do... you can't possibly have opinions, because if you haven't let those opposed to Hitler go to war with him, there is no you – your parents and your entire family will have been wiped out by the Nazis!

I was on a rant – I stood up and raised a glass, and dramatically pointing at my opponent, offered a toast: "He, my friends, is against war! Let's drink to the peacemaker!" In short, I made a scene in polite society, and later my friend told me that at his next birthday he'd be more careful about his seating arrangements.

I am afraid that my friend will have a great deal of trouble with that on his next birthday, because lately, in polite society, not professing pacifism has become an indecent act. Pacifism has grown and spread deep and wide, and shows all the signs of an airborne infection in both the literal and figurative sense.

From 2000 on, pacifists have raised one hysterical cry after another whenever antiterrorist operations are launched.

In 2001, after the attacks on New York and Washington, the American government finally woke up, and resolved to crush the Islamo-Fascist Taliban regime in Afghanistan, which was at the time the very epicenter of evil in the world. Waves of protest rose up; mass demonstrations organized by local activists (on whose time and with whose money, one might ask?) forced many governments to back away from supporting antiterrorist operations in Afghanistan.

I recall one popular television news program featuring a certain gentleman from Massachusetts whose brother had died on September 11th at the hands of terrorists. This gentleman was heading a campaign against the bombings aimed at terrorists and those who harbored them, because innocent people might suffer. As I watched, I suddenly understood. This man whose brother had been murdered by terrorists and the terrorists who had murdered his brother were all part of the same gang. They simply perform a different function: terrorists murder people; pacifists prevent people from defending themselves against the murderers. One should also carefully consider which weighs more heavily in the balance: those who kill outright, or those bent on keeping the victims from defending themselves. It is just such respectable gentlemen that terrorists, laughing up their sleeves, rely on; these gentlemen will certainly never allow any reprisal, they'll twist arms and even legs in Congress to prevent it.

There's been no twisting so far, but I was horrified to see how many advocates our enemies have, and they will twist, eventually. All the passions roused by the current state of Saddam Hussein's Iraq are a sad confirmation of that fact.

The pacifists somehow missed the beginning of military operations to oust Saddam Hussein. Perhaps their initial passivity can be attributed to the massive support shown by the US Congress in 2003, or perhaps to the unanimous opinion voiced by security services around the world that the dictator had built weapons of mass destruction. But now some time has passed, the Fascist regime in Iraq is overthrown, and a country that has never known democracy is clawing its way out of Saddam's totalitarian pit under the constant fire of terrorist groups dispatched from abroad. Just a little more, just a little longer, and this nest of aggression and lawlessness may in fact turn into a normal, civilized nation. We need to help the new Iraq survive, create an Iraqi army capable of protecting its people from these gangs; we just need to hang on until the Iraqis pass a constitution, we need to help them set up free election – just a little more, a little longer...

But oh no – the pacifists are doing everything to prevent that, the pacifists are pressing for withdrawal of British and American troops from Iraq, the pacifists are bent on bringing any progress toward a free Iraq to naught, the pacifists are aiding and abetting the surviving henchman of Saddam Hussein in their quest to return to power.

The assault of American pacifists on the American government is becoming ever wider and ever more implacable. The interests of the nation, the growing threat of Islamo-Fascism, the need for a united front in the war of terrorism are all cast aside in favor of liberal political ambitions thoroughly infected with the leftist disease. "No War!" This meaningless phrase is becoming an obsession for which there is no salvation.

The inevitable loss of life in war, the sorrow of mothers who have lost their sons is being used like a battering ram against any arguments defending the need to continue military operations in Iraq. A mother's grief over a fallen son is beyond measure, but when that grief is trotted out for political purposes, it stops being a natural expression of loss and sorrow, and turns into an illness. This illness is rotting the souls of the politicians who, as their main argument against the war, push the mothers of slain soldiers onto center stage at pacifist demonstrations.

Thus, in our modern world, forces in conflict around the globe are arrayed in the following manner: there is good, there is evil – and there is pacifism. Whose side is <u>it</u> on? Theoretically, on the side of good. But in reality?

Along with the absurd rejection of any and all war, even as a last resort against blatant and implacable evil, pacifism has spawned a vast number of not entirely stable children – the peacemakers.

The wrong-headedness of appeasement comes down to the false belief that an aggressor or a terrorist can be stopped and placated by concessions and peaceful initiatives.

This approach nearly led all of Europe down into a Hitlerian hell. The German Fuehrer, knowing full well how the minds of the cowardly French and English "peacemaker" politicians worked, staged fits of hysteria and rage in order to wring out one concession after another. Once having realized that the "peacemaker's" response would be limited to "peaceful initiatives," he occupied France and most of the rest of Europe without regard to the opinion of these same "peacemakers." Only then did the English wake up, and replace their peacemakers with the War Cabinet of Winston Churchill. This great man's reaction to peacemaking was idiosyncratic – he was the first to just say "no" to negotiations with Hitler, "no" to peace with him no matter what the terms.

The appeasers came up with a devilishly clever saying that always plays back neatly into the devil's hands: "A bad peace is better than a good war." And the average citizen nods, yes, yes, a bad peace is better than a good war, and evil, smirking with contempt, overruns yet another position without a bullet fired.

In our day the peacemakers have taken this thought to its logical absurd: "Peace at any price," or even more boldly, "There is nothing more precious than peace."

In fact, in our lives – both as individuals and members of a society and community – there are a good many things that were, are and always will be more precious than mere peace. There is the future of the family and the nation; there is faith and culture; there is personal freedom; there is the freedom of one's native country. Every one of us might add something to this list, something that is more precious – something that our ancestors were willing to give their lives for.

And nonetheless the average citizen listens and dully bobs his head – no, there's nothing more precious than peace.

For years the appeasers have been tossing food like this to the Islamo-Fascist crocodile in the hopes that, sated, it would ignore them. Now they've seen that it will eat them anyway.

Pacifism is particularly dangerous in that it is superficially attractive, and sometimes thoroughly justified. Many pacifists are kind, sensitive and generally likeable people. When they modestly and sincerely declare, "We're against war," it's hard to say "I'm for it." When they ask sternly, "And would you want your son sent to war?" it's unthinkable to answer "Yes I would." Many outstanding humanists worthy of great respect have been pacifists, and opposing them feels awkward.

Before and during the First World War Albert Einstein sided with the pacifist movement. Like many of his fellow intellectuals, he saw just how senseless and criminal was the slaughter unleashed in the fields of Europe by Europe's own self-serving leaders. But then came the even bloodier Second World War, and this former pacifist, one the greatest scientists on the planet, wrote President Franklin Delano Roosevelt urging him to start work as soon as possible on the manufacture of a weapon capable of killing hundreds of thousands of people in an instant – the atom bomb. This was a difficult step for Einstein to take, but he realized the scale of the evil that had emerged in Nazi Germany, and understood that at this moment remaining a pacifist would be playing into the hands of that evil.

The list of prominent pacifists is so impressive that one can only tip one's hat as they parade by.

But all sentimentality aside, we must acknowledge in all honesty that modern-day pacifism is a mole, a spy in the camp of good, a fifth column in the camp of civilization.

Evil never bothers with any discussion of the inhumanity of war; evil acts according to its own will, and evil attacks when the camp of good is busy debating pacifist ideas.

Therefore – he who seeks peace at any price will inevitably find himself in the middle of a war!

Therefore – he who values peace about all else will lose everything he values!

Because of the disease on the left, pacifism, once one of the noblest components of liberalism, now serves only as an obstruction in the war between good and evil.

And evil knows that, evil appreciates that, evil seeks ways to spread the disease, because it sees here perhaps its only chance to defeat good.

Early in 2006 the Islamo-Fascist fuehrer sent out his most recent message to the world, praising the work done by American pacifists. The fuehrer did not conceal his wish to turn history backward. One would think that pacifists would shudder at praise like this. But no! Our current peacemakers make no effort to hide their alignment to such fuehrers. Soon after the most recent Islamo-Fascist manifesto, the newfound leader of the American pacifist movement, a lady in the grip of the most violent stage of the leftist disease, sought support from none other than the dictator of Venezuela, Hugo Chavez, who hopes to supplant Fidel Castro as the leader of a worldwide alliance against "American imperialism". A photograph of this lady embracing Chavez – a monstrous symbol of the alliance between modern pacifism and world evil – sped round the globe.

Many people say that we are in the midst of a Third World War. Whether one agrees with them or not, it would be foolish to deny that a fierce battle is being waged between good and evil here in the Third Millennium. And even if we don't consider ourselves the Third World War we are at least standing on its doorstep, and our situation mirrors the situation on the eve of the Second World War, when European pacifists obstructed any efforts to rein in Hitler by force.

George Orwell, whose anti-utopian novels vividly depict some of totalitarianism's most grotesque features, wrote in 1942:

> Pacifism is objectively pro-Fascist. This is elementary common sense. If you hamper the war effort of one side you automatically help out that of the other side. Nor is there any way of remaining outside such a war as the present one.

These words were directed at English pacifists in the midst of the Second World War. But God in heaven, how well they apply to pacifists today!

Substitute the modern "Islamo-Fascist" for George Orwell's "fascist" – and you will see a perfect description of the current pacifist movement. One cannot remain on the sidelines; one cannot remain neutral in the worldwide war on terrorism. If modern pacifism opposes all use of force against terror, modern pacifism is on the side of terror. There is no third way.

But you know, reader, what I found most depressing about this entire story is that the mere substitution of one term by another fifty years older perfectly defines the current state of affairs.

That the last fifty years have taught people nothing; they have not cured the pacifists of their ailment. For all the effort exerted, for all the lessons the muse Clio has offered the twentieth century, we see no signs of recovery, or even improvement, in the health of patients suffering from the leftist disease. And that can mean only one thing – that its complications have taken on chronic form.

CHAPTER V

ACCOMPANYING SYMPTOMS: SYMPATHY FOR THE CRIMINAL

> *He who wishes to show mercy should weigh his decision on the scales of his conscience. Will his act of mercy will bring more good to this world, or more evil?*
>
> *The author's opinion*

Another dangerous concept derived from the general theory of nonviolence is the famous notion of "humane treatment of criminals." This too is a symptom of the leftist disease infecting liberalism.

That democratic America should so coddle its serious criminals – terrorists, serial murderers, rapists – makes me sometimes almost physically ill. People who have lost loved ones at the hands of murderers and perverts are forced to hear descriptions of institutions that seem more like health spas than prisons: three nutritious meals a day, showers, televisions, gyms, libraries, sports competitions, activities and the like.

I once happened to read a menu of the meals provided for the captured terrorists now held at the Guantanamo Bay naval base on the island of Cuba: meat, fish, fruits and vegetables, baked goods, cold drinks. In all probability this multinational gaggle of professional killers has never eaten so well as it does now, in an American prison.

However, that's not enough for "human-rights activists." They are demanding that the terrorists be considered prisoners of war and that the rules of the Geneva Convention be applied to them. They have pestered the Pentagon with constant inquiries, humanitarian inspections and complaints about violations of these thugs' rights. The US administration already regrets that it got itself involved in this business with the Afghani terrorists, and would happily let this rabble scatter to the four winds – except that no one wants to take them. And meanwhile the leftists, one step at a time, are achieving their goal: the Pentagon has started building new and more comfortable accommodations for the terrorists – single rooms with an ocean view. According to the announcement that construction would soon begin, the ocean view would alleviate any depression the poor criminals might be experiencing, and also reassure the activists that the prisoners were being treated humanely.

Oh Lord, why did you not direct the feverish energy of these besotted activists toward helping the families of those who perished at the hands of the terrorists? Families that cannot be comforted by an ocean view?

I do not want to be unfairly accused of inhumanity, or of rejecting the idea of compassion or mercy. The quality of mercy is one of the human soul's most remarkable attributes. Help for the needy, compassion for the unfortunate, leniency toward those who've gone astray... without such forms of human kindness our world could not exist. But the leftist disease recognizes no reasonable boundaries. It demands that society show humanity, compassion, sympathy to murderers, to incorrigible thugs and terrorists who have long since ceased to be human. The disease perverts the very notion of human kindness, and turns the war on crime into a farce, and into a mockery of the victims.

For, as it has been said, mercy for the criminal is a crime against the merciful.

To come back to the terrorists being held at Guantanamo, in my view, the US Army performed a great act of mercy in not executing them immediately, at the scene of the crime. All the acts of mercy later exacted by activist groups are an extremely dangerous symptom of the leftist disease.

They are dangerous because eventually they will both encourage terror, and disseminate it.

Have you ever seen a repentant terrorist? I haven't. I have seen frightened terrorists, terrorists asking for mercy, but never a sincerely repentant one. Terrorist obscurantism is irreversible – this is a historical fact. No good, no human kindness, no mercy can cure it or correct it, because the terrorist does not know or understand what human kindness is. It is impossible to explain it to him. American terrorist Timothy McVeigh was explained the difference between right and wrong time and time again – at home, in school, in the military – and then one day he loaded a truck with explosives, parked it in front of a multistory building and blew it up. Hundreds of innocent people, including many children, died.

History teaches us that all attempts at combating blatant evil with kindness and nonviolence will in the end merely foster and disseminate terror. Yet the "sympathy disease" continues to infect.

Another of its symptoms is the wholesale condemnation of the death penalty, no matter what the crime. This is an old and very complex issue.

In late 18th-century Europe people condemned to death were beheaded, or even drawn-and-quartered. In 19th-century Europe they were generally hanged or shot. In the 20th century the preferred method was the firing squad, although particularly heinous criminals – Nazi leaders, for example, – were hanged.

In America, the electric chair was long used, but has since been deemed inhumane, and in a number of states particularly dangerous criminals are executed by injection of lethal drugs combined with a sedative to prevent pain upon death. This was the method used several years ago to painlessly execute the above-mentioned Timothy McVeigh, a monster who murdered several hundred men, women and children. This vampire in human form did not fear such a death; he smugly suggested that they hurry up and put him to sleep.

In short, one cannot help but notice an obvious a historical tendency towards applying the death penalty less often, and making the procedures around it less cruel. In Israel, Europe, and in many other countries the death penalty has been banned outright. Leading humanists have long spoken out against it, most notably Lev Tolstoy and Mahatma Gandhi. Tolstoy linked the issue to the notion of the divine origins of human beings:

> **In each human body there exists a divine spark, and therefore no single person nor any group of people has the right to violate this established union between the divine and the human body, i.e., to deprive a man of life.**

Academician Andrei Sakharov, one of Russia's greatest scientists and humanitarians also spoke out against the death penalty.

I will not repeat or discuss all the usual arguments made by opponents of capital punishment. These arguments are quite well founded, and to some degree convincing. And if we consider the enormous authority that Tolstoy, Gandhi and Sakharov have wielded, one might think that the death penalty itself would be doomed.

Yet for all my great respect for these great pacifists, their arguments against the death penalty look convincing and humane only in the abstract. When a real and quite normal, humane person encounters a real murderer – a bloodthirsty killer who has drowned innocent people, brilliant people (and their families) in the muck of his own death-dealing and perversion, that normal person cries out, "I cannot stand to see this murderer live, I cannot breathe the same air, look at the same sky. For God's sake, for the victims' sake, why should I have to live on the same planet as this monster?"

And I understand that person very well, because there is nothing really humane about the "humane treatment" of a brutal murderer. That sort of humanitarianism is an insult to the victims of crime, and such mercy for criminals is itself a crime against the merciful.

There is no "divine spark" in terrorists who murder children!

In November 2005 a group of suicide bombers from Iraq made their way into a hotel in Amman, the capital of Jordan, and blew themselves up in the midst of a wedding party; dozens of people were killed, and hundreds severely wounded. The explosive device that one of the female Muslim suicide bombers had strapped under her clothing failed to detonate; she was detained by the authorities, and later appeared on television – curious viewers on every continent had the opportunity not only to see a live terrorist in the flesh, but to listen to her revelations. We learned from these revelations that she and her husband entered the hotel together, that she saw that it was a Muslim wedding, that her husband blew himself up in the midst of the crowd, that she also tried to do so, but for some reason the device didn't work. There was not even a glimmer of thought or feeling in the expression on her face. I looked at her and thought, "Well, fine, let's assume this is a rather dull-witted and uneducated woman, she couldn't defy her husband; she went along with the plot. But she's still a woman, and she must have some instinct for good, some maternal instinct, after all. How could she, a Muslim woman, plot to kill people at a Muslim wedding? How could she not remember her own wedding at that moment? Is there really a divine spark in this woman? Is this a woman? Is this even a human being?"

What do we draw from this?

The supposition that the terrorists of who kill innocents today and every day have some measure of the divine spark is an insult to the very notion.

The practical reasons why the death penalty has often has not been applied are well known. Many monstrous crimes have been committed simply because someone at some time demonstrated the requisite mercy to a criminal or, in pursuit of political (or other) goals, refrained from imposing capital punishment. History is full of vicious repeat crimes committed by people who, for "humane" reasons, were not condemned to death.

In 1981 Ayman al-Zawahiri, a member of the Islamic Jihad terrorist group, was arrested for complicity in the assassination of Egyptian president Anwar Sadat. Had he been executed as he deserved, according to law, human history might not have taken the grim turn it has. However, thanks to the political games the new Egyptian president was playing with the extremists, this assassin was instead released from prison after serving a sentence of only three years. In 1990, an Egyptian court finally sentenced al-Zawahiri to death for terrorist activity, but he managed to evade capture. In 1998 he joined Al-Qaeda, becoming its chief ideologue and right-hand man to Osama bin Laden. In 2001 al-Zawahiri organized the terrorist attack on the USA that carried off thousands of lives. American bombers pounded Afghanistan with thousands of pounds of bombs in an attempt to destroy the Al-Qaeda network and kill its leaders. Many people were killed in these bombings, including al-Zawahiri's wife and children, but he again slipped away. Perhaps it might have been better to execute him back in 1981, for his very first murder?

Israel's prisons hold any number of terrorists with the blood of innocents on their hands, but these murderers cannot be put to death. Israel has banned capital punishment. From time to time these murderers are set free – sometimes for political reasons, sometimes in exchange for hostages. Once free, they kill again, and their victims are actually the victims of the ban on capital punishment.

Moreover, one would like to ask opponents of the death penalty, all those great humanists past and present, the following question: "What would you do with an Ayman al-Zawahiri or an Osama bin Laden? Would you continue insist that the same divine spark that once lived in the thousands of innocent men, women and children they have murdered still lives in them?"

In December 2005 there was a curious episode in the history of left-liberal mercy for murderers. In fact it was more than curious – it was outright ludicrous. Public opinion in America literally split into two camps over the Williams case.

Mr. Williams, as the politically correct press so politely calls him, was in fact a cold-blooded murderer. Twenty-some years ago he was the leader of a street gang, and shot four people during a robbery. The court remarked on the particularly cold-blooded and cruel nature of these killings. After the trial, the usual American song and dance began: complaints were lodged and appeals were made on every level, ending with the US Supreme Court. Meanwhile, this monster was spending his days in a virtual resort where, for lack of anything else to do, he began writing edifying children's literature. (American prisons provide wonderful retreats for would-be writers.) Publishers sniffing out sensational stories pounced on Williams's books, while friends and sympathizers rushed to nominate him for the Nobel Peace Prize. The Norwegian Nobel Committee, which had awarded the same prize to professional terrorist Yasser Arafat not long before, felt that it had to consider William's candidacy. The committee at least had the wit not to give him the prize. But that's not the point.

The point is that after twenty years worth of delay and procrastination, after every imaginable court denied Williams clemency because of the particularly heinous nature of his crimes, the time came to carry out the sentence prescribed by law. And then it all started! Heads of leftist organizations, opponents of capital punishment, the press, radio and television all rose up to defend whom against whom? To defend the murderer against his "heartless executioners," meaning not only the state, but the families of the victims and anyone else who demanded the sentence be carried out. Crowds of sympathizers suffering from an advanced and violent stage of the leftist disease mounted hysterical demonstrations in front of the governor's mansion in Sacramento, as California governor Arnold Schwarzenegger had to decide the killer's ultimate fate. I kept thinking in horror that no, the governor would never be able to stand against this mob, that the monster would live to give "life lessons" to our children, and to us. To my surprise, the governor stood fast; after studying the circumstances of the Williams case, he found nothing in them that would warrant leniency. Williams was finally put to death – which prompted another spike in the leftist disease already raging throughout the world.

In the governor's home city of Grac in Austria, the Schwarzenegger Stadium was hastily renamed, and any references to Grac's famous native son disappeared – so sad were his fellow Austrians that the gangster was no longer among the living. All of Europe, in one voice, castigated Schwarzenegger for siding with the law rather than with an ochlocratic mob of quasi-humanists. Here in America Schwarzenegger has been termed no more and no less than "a cold-blooded murderer." Take note here: it's not the man executed for killing four innocent people, but the governor who upheld the law of the land who is deemed a cold-blooded murderer. This loss of any sense of justice or fairness is yet another effect of this violent stage of the illness.

Apropos of the violent stage of the leftist disease, the mass insanity that presents itself as aggressive ochlocratic demonstrations is a direct result of the proliferation of the leftist bacilla by prominent public and political figures both in Europe and America. Since we are bound by the principle of medical confidentiality, we cannot divulge any names. Nevertheless, everyone knows who they are. When a respected senators, congressmen, congresswomen, politicians, journalists, artists suddenly start flapping their arms wildly, shouting out loud, incoherent slogans, and rolling their eyes, I am overcome by a feeling of inconsolable sadness – here's another one for the locked ward.

To conclude, let me relate one final story of false mercy, one of the most horrible stories that illustrate just how degraded humanism and liberalism have become in our time.

In February 2005, in a small town in the state of Florida, a charming little nine-year-old girl by the name of Jessica Lunsford disappeared, snatched out of her own bed in her parents' home. It did not take the local police long to focus on a prime suspect, because in this was one place in which the names of sex offenders went onto a registry, and were generally well known. By March this piece of scum (I will not say his name, because beasts like this don't deserve human names) was captured at the home of relatives in Georgia. He soon confessed that he had crept into the Lundsfords' house in the middle of the night, pulled Jessica out of bed, half-strangling her in the process, and then dragged her off to the woods, raped her, killed her, and buried her body. I'll dispense with any further detail.

The murdered girl's body was soon found. At her funeral service in a local church, the pastor exhorted the congregation to show mercy and compassion, and to forgive.

Some might call this just another sex crime by a madman – there are thousands such every day, all over the world. And they might ask as well – what does this have to do with liberalism?

My answer is that while it has nothing to do with liberalism itself, it has everything to do with the disease from which liberalism currently suffers.

After reading about the pastor's request, I grew curious about the murderer's past: maybe he hadn't been such a bad kid at first, maybe he'd just snapped. Amidst the hundreds of articles on Jessica's murder, her funeral, her parents and relatives, information on the murderer himself had been virtually lost. His "heroic" past seemed to have been carefully tucked away like some shameful secret our society wished to hide. Finally, however, in one article about the crime, I found a brief resume of this hero of our time, a biography, so to speak.

I was stunned.

Childhood aside, in the roughly thirty years of our hero's life before his murder of Jessica he was arrested 24 times (twenty-four!) and just as many times released – that is, once a year, on average. One could have studied the criminal code just on the basis of this animal's record, because there seemed to be no crime he hadn't committed. His resume includes armed robbery, breaking and entering theft, nonpayment of debts, resisting arrest, illegal weapons possession, use and distribution of narcotics, lewd and lascivious behavior, fraud, creating a public disturbance while under the influence of drugs or alcohol, frequent refusal to appear in court – all this besides being a registered sex offender. During two of his robberies he had attempted to have sex with underage girls. A little work, a little play? Jessica's murder was the natural finale to this freak's criminal career; one needn't be a professional criminologist to realize that sooner or later something like this was bound to happen.

Any normal person reading this murderer's record should be consumed with righteous indignation.

Why was this monster at liberty? Why was he freed time and time again? Who let this happen? Why? Why is not a single one of those people who set him free time and time again not sitting with him in the dock? Who will ever answer for the tragic death of a nine-year-old girl? Who will answer for her parents' sorrow?

No one, I tell you, because it was mercy, it was sympathy for the criminal that set him free. Sympathy for the criminal is what killed Jessica Lunsford. The leftist disease has twisted the minds of those who could have and should have isolated this scum from the rest of society.

The degradation of liberalism brought about by pseudo-mercy for criminals has rendered society helpless, because tomorrow the next monster freed on ideal "humane" grounds might well break into your house, or mine. No – it's not just that he might. He will. He has nothing to fear from this merciful and compassionate society.

Do you know the most terrifying thing about the story of poor Jessica Lunsford?

The fact that the pastor preached forgiveness for her murderer.

And that no one shuddered when he did!

CHAPTER VI

ONGOING CONDITION:
OBSESSION WITH SOCIAL
DEPENDENCY

This ain't the welfare office!

Popular saying

Welfare from the cradle to the grave!

Popular wish

The burden of the welfare state is high indeed, both in economic terms and from the perspective of human dignity.

Nils Karlson

Europe has always been haunted by all sorts of specters that, unfortunately, materialize in the very worst forms imaginable. Such was the case with Communism; such was the case with Fascism. In our time another specter is haunting Europe – the specter of socialism.

Granted, this specter is not the Marxist-Socialist one that was eventually embodied in the gigantic network of slave-labor camps built in the Soviet Union. This specter came into the world in the mid-twentieth century. Its youthful father was the new liberalism; its aging mother was the eternal idea of socialization/nationalization of property.

In our day this specter has acquired both a shape and a name. In scholarly literature it is called "European socialism."

Wait a minute! Stop! God forbid that anyone think that this author intends to discuss, let alone criticize, socialism here. No, this author cannot afford to take such liberties, and therefore must boldly state that socialism is too broad a topic to even attempt to cover in this particular work.

The topic of socialism comes up only because its foundations (which are themselves controversial) serve as host to one of the most dangerous and parasitic manifestations of the leftist disease: the obsession with social dependence.

The English name for this particular symptom is short and expressive. Welfare. This word is generally translated into Russian as "a government social assistance program." But most Russian-speaking immigrants don't bother with such lengthy translations and have boldly introduced the word "velfer" into their Russian lexicon.

In Europe, the welfare symptom can be observed wherever and whenever European-socialist states are under construction. Avoiding the term "socialist state," the classic theoreticians of the new liberalism have introduced instead the notion of "the welfare state," a definition which in our view throws this symptom into high relief.

In Europe the construction of welfare states proceeds apace, and perhaps the best illustration of this is what has happened in Sweden.

The fact is that the Swedish welfare state is ubiquitously held up as the perfect model not only for Europe, but for the entire world. There are two reasons for this. On the one hand, the Swedish welfare system is the oldest in Western Europe, and the transition to government patronage has gone further and deeper than in any other Western nation. On the other hand, the Swedish welfare state has enjoyed a certain success: in 2004

Sweden's gross domestic product (GDP) per capita ranked eighth in the world (just behind the United States, and ahead of many other developed nations).

But in order to understand the Swedish welfare phenomenon, we must make at least a brief historical digression.

Between the late nineteenth century and the mid-twentieth, Sweden was developing along classic capitalist lines. The development was rapid, to say the least; Sweden was outpacing every other country in the world. In fifty years time, this poor, backward country on the outskirts of Europe had transformed itself into a modern nation at the forefront of science and industry.

What is important to our discussion is that during this period of rapid capitalist development, Sweden had the lowest taxes of any developed nation, and also the lowest level of government intervention in the economy or the private lives of its citizens. In 1950 Swedish taxes constituted 20% of the GDP, in comparison to 24% in the US and 27% (on average) in Europe. Swedish capitalism was producing impressive results with minimal government involvement and minimal tax rates.

However, with the rapid spread and progression of the leftist disease in the late 1950s and early 1960s, and with the oldest social-democratic labor party in Europe at the helm, Sweden abruptly changed political and economic course. It wanted to demonstrate to the world that, on the basis of new-liberal principles like socialization and government regulation that there could be a so-called "third way" societies could develop, a path that lay somewhere between American-style capitalism and Soviet-style socialism – a path toward a welfare state.

Today the Swedish welfare state is a finished product, and we can make some sort of assessment of its success. There is a great deal of literature on this topic, but I was looking for the professional opinion of a credible Swedish expert who understood the issue from the inside. I found the sort of evaluation I was looking for in a series of articles by Professor Nils Karlson, which have been published under the general rubric of *"European Socialism"* in The Architecture of Modern Political Power (AMPP).

Nils Karlson is the president and general director of the Ratio Institute, a research center in Stockholm that focuses on political science and economics. Karlson formerly served as president of Stockholm University; he currently teaches at the University of Uppsala. His assessment of Sweden's welfare experiment is quite negative:

> For many years the Swedish welfare state portrayed itself as a model to the world... a triumph of modern civilization... However, the reality of the Swedish model has become quite different to what was intended and to what many people believe to still be the case. Five stylized facts may illustrate the present situation:
>
> I.   No job on net have been produced in the private sector since 1950.
>
> II. Non of the top 50 companies on the Stockholm stock exchange has been started since 1970.
>
> III. Sweden has dropped from fourth to 14th place in 2002 among the OECD countries in terms of GDP per capita since 1970.
>
> IV. Well over one million people out of a work force of around five million do not work in 2003 but live on various kinds of public welfare programmes such as pre-pension schemes.
>
> V.   A majority of the adult population are either employed by the state or clients of the state in the sense that they have a majority of the income coming from public subsidies.

Karlson also goes on to note that the establishment of a welfare state inevitably means ever increasing taxes. At the beginning of the twenty first century, taxes in Sweden constituted 53% of the national GDP, more than in any other country in the world. In an article published in 2004, Karlson writes:

> An average Swedish worker pays 60 percent of this income in taxes, if direct and indirect taxes as well as social security contributions are summed up. In a similar manner, holders of shares of companies on the Stockholm Stock Exchange pay around 60 percent in taxes if company taxes, property taxes, taxes on dividends and so on are summed up. Successful entrepreneurs sometimes pay even more. This largely explains the five stylized facts reported above.

In the last half-century Sweden, once a country with the lowest taxes in the world, now has some of the highest, and there is no sign that this trend is about to reverse itself. Each new publicly funded entitlement creates a new category of dependents and uninsured. This, in turn, leads to a drop in manufacturing, to job loss, and to the creation of yet another category of dependents, which in turn requires that the welfare state introduce new forms of support and new tax increases. So far it has been impossible to break the vicious cycle: tax increases – inefficient economy – tax increases.

In his study of the Swedish model, Nils Karlson came to the conclusion that the consequences of socialization/nationalization cannot be measured in economic terms alone. The welfare state inevitably deforms and degrades both character and morals.

For those who experienced socialism Soviet-style, this conclusion is neither new nor surprising. The main effect of the Soviet socialist experiment was not the catastrophic drop in economic efficiency that the country suffered, but the utter degradation of personal and public morality. Decades of socialism have bred dependence, irresponsibility, and an unwillingness to do honest work, and the former republics and satellites of the USSR are now reaping the results.

Nils Karlson has scrupulously examined the influence of the welfare state on human dignity, which he defines as "the personal responsibility of every individual for his own life… with equal respect for others' liberty." A healthy, able-bodied person who lives off government subsidies cannot possibly maintain the same level of dignity as the free individual, and "any Swedes," he writes, "have become heavily dependent on the state and have neither means nor the ability to take responsibility for their own lives." Karlson provides an interesting chart showing the relationship between dignity and taxes, which demonstrates that at a certain level, tax increases precipitate a sharp drop in the average sense of self-esteem and dignity.

Summarizing his research on the Swedish model of the welfare state, Nils Karlson writes that the burden of the welfare state weighs heavily both on the economy and on human dignity, and if we want to make the economy more efficient, and to promote human dignity, it is imperative that we make radical reductions in social assistance programs funded by the government; i.e., significantly reduce the level of government intervention both in the economy and in private life.

Readers may find themselves perplexed: "If Professor Karlson is right, and welfare states don't work, how is it that Sweden maintains such a high standard of living?" While this question lies absolutely outside the thematic scope of the present article, my own point of view is the following.

There are many reasons for Sweden's relatively high standard of living. One of them, for example, is the fact that Sweden did not take part in either world war; that for two hundred years the country has enjoyed peace and prosperity, and thus happily managed to avoid the twentieth-century catastrophes that other parts of Europe did not. Another significant factor is Sweden's relatively small and stable population; it is much harder to maintain a high standard of living in more populous countries that experience significant migration. We might also cite the high level of education traditional for the Swedes, their northern stolidity and thoroughness, their native industry and so on.

In my view, the chief reason why Sweden has maintained a relatively high standard of living is that the Swedish welfare experiment was launched from what was already a very high platform; it began at the socio-economic level that the country had reached in the early 1960s. This very high level was in its turn a result of the half century of rapid economic growth enjoyed by Swedish capitalism, which, as we have already noted, took place in a uniquely peaceful and prosperous environment.

In this sense the Swedish welfare experiment is strikingly different from the well-known attempts to build socialism in countries where the starting point was much lower on the economic scale, where the capitalist manufacturing base and free markets were either underdeveloped, or not developed at all. Here we cannot help but marvel at the perspicacity of Karl Marx, who insisted that socialism could be built only in countries that had reached an advanced stage of capitalist development. Soviet Marxists took the opposite tack, and attempted to build socialism in a relatively undeveloped country; naturally, nothing came of that except an enormous slave-labor camp and an extraordinarily low standard of living.

Thus what we see in today's Sweden versus "what people still believe to be the case," as Nils Karlson put it, are the vestiges of what Swedish capitalism once achieved. The Swedish welfare state, in its eagerness to develop an ever more paternal relationship to its citizens and to inculcate in them a sense of dependence on the state, is a parasite feeding off the achievements of its own capitalist past. In rejecting its own native capitalism, the Swedish welfare state is, as they say, sawing off the branch it's sitting on. All of this threatens to lead to dead-end, socialist deterioration unless the government moves to radically cut social welfare programs. Nils Karlson warns that "other countries, such as Germany, which are approaching the Swedish situation, should beware. No one knows for sure when the point of no return is reached."

It seems to me that the point of no return is reached when dependence on society becomes chronic, and those suffering from it begin to chant "Welfare for all from the cradle to the grave!" The schizophrenic nature of this slogan should not blunt the vigilance of people of common sense: not only Sweden, but other European countries are moving to make it a reality. And Canada is not far behind.

These persistent notions of dependence on society are less evident in the United States than in Europe; here, the idea of private enterprise free of government controls is still highly popular; the idea of capitalism itself is virtually sacred. Nonetheless, even here we cannot help but see the beginnings of a welfare state.

Mark Zaltzberg, a professor at the University of Houston, recently published an Internet article entitled *"Goodbye America."* This is only one of many articles in which the alarming symptoms of a national slide into a pit of social dependency are discussed. What is interesting about this article is its author's total dismissal of any sort of political correctness. He writes:

> After living 23 years in the US and the previous 45 in the USSR, I have begun to note with horror the striking resemblance between socio-political trends in both countries. Over the last three or four years it has become clear to me that America is building communism without even realizing it... What happened suddenly in Russia in 1917 is what is now happening gradually in America... the result is that both houses of Congress, state legislatures and various other government bodies are full of people whose only goal is the only too familiar call "Take it all away; divide it up."

It's possible that the author is exaggerating somewhat, since we haven't quite gotten to "taking it all away." But the idea of dividing everything up already does exist: this is "the welfare syndrome." Politicians and public figures suffering from it regularly dun the government, (in essence, American taxpayers) for one new appropriation after another, all to pay for benefits and subsidies for people who either don't know how to work, work very little, or don't work at all.

I repeat for those of you who may not have understood: we are not talking about help for the old or the infirm; we are not talking about pensions for retirees or support for the disabled. We are talking about financial support for shameless freeloaders who prefer an irresponsible, passive life on welfare to a perhaps harder but more honest way of life. Frankly, I do not believe that, in America, a healthy, relatively young person cannot find work. Of course, it might not be a job in his own field, especially if it is an intellectual one, but to find no work at all? Impossible. Nonetheless, the number of people who prefer to stay on welfare is growing at a frightening rate – and this in a country, please note, built by generation after generation of determined and ambitious individualists who were willing to work to exhaustion in order to lead a worthy life, to gain fame and fortune.

Dear reader, please allow us to digress just a little from our topic, and to give you some relief from the well-meaning madness of welfare.

In autumn 2005, I happened to be in California, a place from which America has often challenged the rest of the world, challenged nature, challenged the forces of ignorance and stagnation.

In America it's often said that many great things have come out of California garages. The most historic garage of all is located at 367 Addison Avenue in Palo Alto, CA. There's an inscription on it, reading simply "Birthplace of Silicon Valley." No further explanation is necessary; great things don't need long descriptions.

In 1939, in this space totaling twenty square meters, two young engineers named Bill Hewlett and Dave Packard founded their firm. In the beginning, they <u>were</u> the firm; they were its owners, founders, and only employees; they started without pay, bonuses or benefits. Their start-up capital came to a grand total of $538 dollars, all borrowed. Sixty-five years later HP (Hewlett-Packard) had over 100 thousand employees, and its worth was estimated at fifty billion dollars. These figures, however, tell us little about the true nature of the revolution begun by these two bold young men in America's most important garage. These pioneers of Silicon Valley drew a new map of corporate relations and corporate culture; they introduced methods and approaches that remain the stuff of legend in the business world. For over half a century the firm was a virtual utopia, an island-state made prosperous not by government welfare but rather by the inspiration and perseverance of its own citizens.

Curiously enough, Bill and Dave's first customer was Walt Disney, who bought eight audio oscillators that the young engineers had designed and built with their own hands. Disney too was a pioneer, working out of another now famous California garage. Walt Disney's cartoons have conquered the world; the entertainment empire he built is unmatched in scale or creative power. And yet Walt Disney began just as Hewlett and Packard had; he got his start in his uncle's garage in Los Angeles, with a stake of just few hundred dollars between himself and his brother Roy, and the rest borrowed from relatives.

Perhaps, in these sunny hills on the shores of a mighty ocean, it was nature itself that brought out the urge to try, to dare, to seek fame and fortune, to challenge the forces of mediocrity and ignorance.

It was from here that those who had already caught gold fever headed north – some to their deaths, some to fame and fortune

– rascals and rogues, brave and foolish alike. Those who survived came back to build the great cities of California.

Here, in Hollywood, tough, mercenary producers and ambitious, hitherto unknown actors broke into the international film industry.

Here, in San Francisco, at the turn of the 19th century, there lived a young man who had no education to speak of, and no one to count on but himself. He never turned down a hard or a dirty job; in his free time, late into the night, he read voraciously; he wrote stories and novels. The young man dreamed of becoming a great writer, but found no one willing to publish him, even for free. At the age of twenty-one he threw down a final challenge to fate, and went north to Klondike for two years, to prospect for gold. What he found in the frozen Klondike was something more precious than gold. He found worldwide fame. When he returned to San Francisco with new stories to sell, publishers fought tooth-and-nail for the rights to them, and by the age of twenty-eight Jack London was the most popular and highly-paid writer in America.

Bill Hewlett and Dave Packard, Walt Disney, and Jack London represent different professions, different eras, and different generations, but what they have in common is not only talent, but dignity – the dignity of free people who expect no handouts from the state.

How would the lives of people like this played out had they lived in a welfare state? Would they have fought hard to make a decent life for themselves, or would they have preferred the cheerless but safe prospect life on the dole? It's hard to imagine them choosing the latter.

In a welfare state, would Silicon Valley or Hollywood ever have come to be at? At a sixty-percent tax rate on business, would their founding fathers have even tried to create anything at all? I doubt it.

Today's California looks rather sad in comparison.

For example, there is the strange and ubiquitous phenomenon of the happily unemployed and homeless. I encountered them in parks, on the streets of San Francisco, on the beaches of Los Angeles.

On one of the main streets of San Francisco, I happened to see three homeless people setting up camp for the night directly across the street from my hotel window. This colorful trio consisted of an old man and a young couple. They settled down under an awning, spread out some sort of blanket on the sidewalk, sat down, and began an animated discussion. From time to time the woman would disappear, then return with some food.

In Santa Monica, in a beautiful park overlooking the Pacific Ocean, in the middle of the day, I saw people sleeping, or simply basking like seals in the California sun. They were all fairly young, and next to each of them stood a grocery cart filled with their belongings. These were the California homeless. They do not want to work, I was told; they do not want to live in houses. They are clothed and fed by charitable organizations. A number of them have mental problems, but the majority simply like this way of life.

At magnificent Redondo Beach I watched the morning wake-up call. Young men rose from the benches where they'd spent the night to make their leisurely way to the entrance, where local activists were handing out sandwiches and hot coffee. I was told that they also served hot soup at midday, and fried chicken with vegetables for supper.

One can see scenes like this in almost every major city in America. From time to time, people may complain that these panhandlers are making life miserable for everyone around, but no one seems to want to do anything about it. The obsessive urge to rely entirely on the state afflicts both those who take and those who give.

But America's parasitic homeless are merely the tip of the iceberg; the bulk of it is made up of a crowd of perfectly able-bodied people who try their best to grab whatever they can from their government, their union, their workplace, their insurance company, even from other private individuals. To get a million dollars from a tobacco company because they ignored the warning printed on every pack of cigarettes, and smoked themselves into lung cancer. A million from a restaurant or a grocery store because they ignored the basic rules of nutrition and ate themselves into a wheelchair. A hundred thousand from a cousin because they slipped and fell on her porch. Half a million from a friend who made the mistake of offering them a ride home and got into a traffic accident. A million from an obstetrician years after a child is born, because the child doesn't think very clearly.

The advertisements are everywhere: contact this or that law firm; we'll help you get the compensation you deserve for... well, for whatever, just bring us your claim. America easily leads the world in the production of lawyers. This is supply and demand at work.

Once a nation of risk-taking, creative individualists, America has become a haven for extortionists.

In place of the legendary American dream – starting with nothing, making it on one's own, taking risks, bringing all one's strengths and abilities to bear in order to win fame and fortune – we have either petty litigation or a calm, passive existence paid for by the government or the unions.

Professor Zaltzberg asks:

> What has happened to the American hero, the staunch, hardworking individualist who relies on no one but himself? The hero who works to exhaustion, who knows that only the strongest will survive? After all, these are the people who once created the richest and most powerful nation in the world. These are the people who for over two hundred years came to America from every continent on this earth, knowing perfectly well that this wasn't "the welfare office." They relied on no one but themselves. Alas, Jack London's heroes, and Jack London himself, have slipped away, into the black hole of the welfare state.

And indeed, where is that once heroic American? Where has he gone? Probably, he's hanging out in some park in Los Angeles, panhandling on the streets of New York, or else suing for all those millions that will let him to quit his job and never work another day in his life.

In Biblical story of Adam, the first man, we read that the Lord God "sent him forth from the garden of Eden, to till the ground from whence he was taken."

And God told Adam:

> "In the sweat of thy face shalt thou eat bread, till thou return unto the ground; for out of it was thou taken."

Translated from lofty Biblical language into everyday, ordinary terms, this injunction might be understood as follows:

> Folks, this ain't the welfare office. You have to work for a living on the good planet Earth.

Nonetheless, the number of people who are ignoring this obligation at others' expense is growing by leap and bounds. The welfare syndrome associated with the leftist disease makes it easy for them to feel comfortable about being dependent on society, because on every step of the way politicians also suffering from the disease pound into them that everyone else is obligated to support them, and that they bear no responsibility for their own lives.

But, as Nils Karlson has said, "No one knows for sure when that point of no return is reached" – the point at which quantity becomes quality. And that is when human civilization will crumble under the unbearable weight of welfare for one and all – because there will be too few creative or simply hardworking people left who are ready and willing to take on the hard work that the Lord God has assigned us.

CHAPTER VII

MAJOR COMPLICATIONS:
THE ANTI-ZIONIST STAGE OF
JUDOPHOBIA

> *My friend... when people criticize Zionists,*
> *they mean Jews. They are talking anti-*
> *Semitism.*
>
> *Martin Luther King*

> *I have a premonition that will not leave me;*
> *as it goes with Israel so will it go with all of us.*
> *Should Israel perish the holocaust will be upon*
> *us.*
>
> *Eric Hoffer*

It would seem to be an axiom set in stone: liberalism and anti-Semitism are incompatible.

And indeed, equality before the law, tolerance of a variety of faiths and cultures, fairness and justice in international relations are the basic principles and categories that have always underlain classic liberalism. It was the late nineteenth-century liberal movement that brought about the emancipation of the Jews of Europe. To this very day, most Jews in the United States, and the few remaining in Europe, tend to support liberal political parties.

But nonetheless, the leftist disease is just that – a disease. First it breaks down healthy principles, and as the disease progresses and enters into a critical stage, it turns those principles into ulcerous growths that destroy any healthy tissue remaining.

Probably, first off, we should note that liberal movements firmly reject the blatant Judophobia cultivated by right-wing extremist neo-Fascist groups. For a true liberal, even an infected one, there is nothing worse than being taken for an anti-Semite.

But the leftist disease shows no mercy, even for the purest and most honorable of people, and has found a convenient breach in the liberal wall; it has invaded the once healthy organism called liberalism and has metastasized into a malignant tumor called Judophobia.

The breach? Anti-Zionism.

"Please!" says the wide-eyed liberal, blinking in surprise. "I absolutely reject any and all forms of anti-Semitism. If I don't accept Zionist ideology, that has nothing to do with Jews. Since when is criticism of Israel's Zionist politics automatically anti-Semitic?"

This is no more than a shell game; the wide-eyed, concerned activist deftly substitutes his rejection of Israel's right to exist for his undeniable right to criticize Israeli policy. And this con artist assumes that the fools in the audience will fall for his trick. However, not everyone in the audience is a fool, and fortunately, the trick doesn't always work.

Anti-Zionism and anti-Semitism are two sides of the same coin, and it is impossible to be one without being the other. The great American religious and political leader Martin Luther King came to the same conclusion:

> The anti-Semite rejoices at any opportunity to vent his
> malice. The times have made it unpopular, in the West,

to proclaim openly a hatred of the Jews. This being the case, the anti-Semite must constantly seek new forms and forums for his poison. How he must revel in the new masquerade! He does not hate the Jews; he is just anti-Zionist! ...

My friend, I do not accuse you of deliberate anti-Semitism. I know you feel, as I do, a deep love of truth and justice and a revulsion for racism, prejudice, and discrimination. But I know you have been misled – as others have been – into thinking you can be 'anti-Zionist' and yet remain true to these heartfelt principles that you and I share. Let my words echo in the depths of your soul: When people criticize Zionism, they mean Jews – make no mistake about it.

This strict correlation between anti-Semitism and anti-Zionism is an indisputable law of history, which we have dubbed "the law of Martin Luther King." This law exposes what left-liberals always try so hard to conceal; namely that anti-Zionism is simply a means of masking anti-Semitism. Martin Luther King was amazingly apt in his choice of the word "masquerade" to describe such attempts at camouflage.

In our day, the Judophobic masquerade has become a highly sophisticated show featuring a variety of masks meant to represent humanism, justice, peace, and democracy.

One mask is that of "the Middle-East peacemaker;" another is "opposition to Israeli aggression;" a third is "humane treatment of Palestinian terrorists;" a fourth is "the democratic Israeli-Arab state;" a fifth is "concern for Muslim shrines" on the Temple Mount in Jerusalem. Liberalism has reached the height of sophistication here. But all this sophistication can be reduced to one simple thing – wiping the state of Israel off the face of the earth.

In left-liberal circles in Europe, fits of Judophobia are becoming increasingly violent.

In early January 2006 the Norwegian minister of finance, who also heads the Socialist Left Party (what a name, ladies and gentlemen!) called upon her fellow Norwegians show solidarity with the Palestinian cause by boycotting products from Israel. "I no longer buy products made in Israel," the high-ranking lady declared proudly, displaying her malignant sores for the entire world to see.

Interestingly enough, not long before this declaration, the Israeli government had moved Israeli settlers out of Gaza by force, and had handed the land over to the Palestinians. The latter responded by mortar attacks on Israeli towns. Common sense would dictate that the Norwegian minister of finance might have at least chided the Palestinians for their reaction. But that's common sense... and here we have a woman who has lost her senses entirely. Note that she does not condemn the recruitment of young Palestinian women as suicide bombers, or the use of suicide bombs to kill Israeli women and children. Her skewed sense of right and wrong tells her to do everything in her power to strangle the economy of a country that is trying to protect its citizens.

Nor do high-ranking left-liberals in the US want to be left in the dust by their European counterparts. On 22 January 2006, at a conference in Herzliya, a certain former American president who in his day stood by and watched as Iran turned into an Islamo-Fascist theocracy, declared that the chief obstacle to peaceful resolution of the Arab-Israeli conflict was what? – Jewish settlements. Please note! The main obstacle to peace is not Palestinian terror attacks, not Palestinian Authority mob rule, not the refusal of Palestinians to recognize Israel's right to exist, but Jewish settlements. And this, I repeat, was said <u>after</u> the unilateral decision to remove Jewish settlers from Gaza and <u>after</u> the number of terror attacks had increased.

Another ever more popular guise is the condemnation of Israel's nuclear weapons policy.

Lately, a former TV journalist has been popping up quite often on American television screens. Once upon a time naive Soviet dissidents rather liked him; now, his illness has progressed to the point that he is hardly recognizable. His performances tend to follow a monotonous anti-war script that generally ends with him shouting his lungs out: he usually starts with a routine and relatively calm statement condemning "the criminal aggression perpetrated by the US in Iraq," but then his eyes suddenly begin to bulge, his face turns red, his mouth twists in hate, and he almost screams out his rhetorical question: "Why is it that Israel can have nuclear weapons, but Iraq and Iran cannot?"

Ordinary viewers who unthinkingly and automatically trust "the peacemakers," bounce.

Please, dear reader, spare me the necessity of resorting to banalities; I have no intention of answering the viewer's question. If the person asking this is simply masking his true intentions, i.e. a wish to see Israel defenseless against a Muslim world numbering in the billions, then there is nothing to talk about. If he sincerely does not understand why the tiny democracy of Israel can and must have weapons of containment, and why the sea of totalitarian regimes surrounding it should not, it's a purely medical issue.

Curiously, the question has brought liberals into line with the views of ultra-reactionary military groups.

On 1 December 2005, General Yuri Baluevsky, head of Russia's General Staff, stated the following:

> The United States demand full transparency from North Korea on the latter's nuclear programs, but shut their eyes to the fact (which has been confirmed by experts) that Israel has long had an impressive arsenal of nuclear weapons.

The utter cynicism of this comparison between a totalitarian regime, North Korea, one of the main supports of the axis of world evil, and democratic Israel, which lives with the constant threat of annihilation at the hands of another part of the axis, Islamo-Fascist Iran, tells us something about this general's intellectual level. But for our purposes, that is not the point. General Baluevsky is hardly likely to be infected with the leftist disease, and even less likely to harbor liberal ideas. If he were, I would not have mentioned his name. Nonetheless, he is saying exactly the same thing that left-liberal politicians in Europe are saying; or rather, the politicians suffering from this ailment are parroting what comes naturally to the general.

This particular manifestation of the leftist disease is all the more acute in the light of the Iranian ayatollahs' nuclear ambitions. Iran, which has openly announced its intention to "wipe the Zionist state from the face of the earth," indignantly poses the same question at every opportunity: "Why is it that Israel can have the atomic bomb, but Iran cannot?" In September 2005, this question was discussed at a meeting between Israeli legislators and American senators and congressmen. The Israelis were disappointed in their American colleagues rather phlegmatic reaction to the Iranian threat: ignorance may be bliss for certain circles in Congress, but ignorance is a luxury that Israel, given its history, cannot afford. Tomi Lapid, head of the opposition party in the Knesset, had the following expressive comment:

> The Jewish nation lost six million people because the West did not understand who Hitler was... We are not prepared to sacrifice another six million lives because the world does not understand who the ayatollahs are.

A brilliant assessment of the situation – and if we add that all of this would-be confusion was and is a direct consequence of the disease that has invaded and perverted liberal ideals, everything falls into place.

Another guise the disease may take is the advocacy of "a democratic and unified Arab-Jewish state in Palestine." What, many say, is wrong with Arabs and Jews sharing a common state, Palestine, with equal rights for citizens to settle wherever they like?

The ingenious idea of the "peaceful" destruction of Israel was hatched back in the 1970s by the Soviet KGB, which understood much earlier than its Arab client-states that Israel could not be defeated by conventional warfare. Once efforts to "drive the Jews into the sea" failed, extremist Arab regimes took that idea to heart. In our day, the idea of creating a bi-national Arab-Jewish state in place of Israel has become a plank in the official platform of the League of Arab Nations, under the rubric of "peaceful resolution" of the conflict in the Middle East. What an amazing and odd metamorphosis! An organization that has spent the last fifty years saying "no": "no" to the existence of Israel, "no" to negotiations with Israel, "no" to peace with Israel, is suddenly, touchingly, concerned that Jews should live in peace. Why?

One effect of the leftist disorder is that liberals diffuse toward the most blatant manifestations of evil. That is precisely what happened in this case: at some point, the left-liberals borrowed the idea of a single, democratic Arab-Israeli state from their predecessors, the KGB and the Arab League. How could they possibly resist such a tempting and simple way to destroy the loathsome Jewish national liberation movement: the expulsion of the Jews by means of gradual, regulated Arab immigration.

If, however, some honest liberal truly and sincerely does not understand what would be so bad about establishing a bi-national state in Palestine, the only thing I can advise is to see a psychiatrist – <u>now</u>. The disorder is becoming chronic, and if the patient does not take immediate steps, it will soon become terminal.

Curiously enough, the idea that Israel and the Palestinian Authority should be replaced by a single, two-nation state has found supporters even among Jews, which fact very much delights hard-core Judophobes. But there is nothing terribly surprising about this; the leftist disease does not recognize ethnic boundaries, and strikes liberals at random, whatever their ethnic background.

The disease so blinds the liberals that they see no danger in anti-Zionism at all: no danger to Israel, no danger to the West, no danger to liberalism itself – all this despite the opposing side's clear statement of its purpose and strategy.

In October 2005, Mahmoud Ahmadinejad, a longtime terrorist, made the following statement at an international conference called "A World without Zionism":

> Israel must be wiped off the map of the world, as our teacher the Imam Khomeini has said. The creation of the state of Israel was an act of international aggression against Islam. The Islamic world will not allow its historical enemy to exist within its borders.

Here we are on the threshold of the 21st century – and we're suddenly back in the Dark Ages?!!

America, France, Great Britain, Germany, Canada, Spain, Russia and even the UN rather feebly scolded the cannibal for his impolitic declaration. They had to react somehow, but what if the thug pulled a knife? Very scary. "Propagandistic rhetoric is counterproductive in a volatile region like the Middle East," said a statement released by the Russian Foreign Ministry. And that's it? An unequivocally expressed intention to annihilate an entire people in the name of Islamic ambitions is not a call to genocide, but merely "propagandistic rhetoric" that is "counterproductive" to something or other?

Well, and how did the liberally inclined religious, humanitarian and human rights organizations who are so concerned about violations of the rights of terrorists and uncomfortable prison accommodations react to his statement? Not one of them condemned the bloodthirsty proposal by this medieval obscurantist with an atom bomb up his sleeve!

Unlike the liberals blinded by the disease afflicting them, the president of Iran clearly sees the chain of cause-and-effect:

A world without Zionism is a world without the Jewish people.

A world without the Jewish people is a world without Judaeo-Christian civilization.

A world without Judaeo-Christian civilization is a world without Christianity.

A world without Christianity is the strategic objective of radical Islamo-Fascism.

No decent person would find any link in this chain acceptable. But the devil always tries to forge his chains whole.

In the early part of the 20th century, the prominent Russian philosopher Nikolai Berdyaev warned against that same chain of logic; the only difference was that he was talking about German Fascism and Soviet Communism.

In the latter half of the 20th century, as Islamo-Fascist ideology began to take shape, the prominent American writer and sociologist Eric Hoffer, whose words serve as one of the epigraphs to this chapter, clearly saw how catastrophic the consequences of such a logical sequence of events would be.

Those suffering from the leftist disease do not see this chain at all, but some individual links are quite to their liking. The disease does not only cause blindness, which is bad enough; it also leads to a psychological aberration whereby white is perceived as black, and vice versa.

Herein lies the chief danger, because if we cross all our t's and dot all our i's, and also note without mincing words that the president of Iran has only one way to achieve his strategic objective. That way is a complication of the leftist disease. Barring this complication, the ayatollahs have no chance of success, even if the entire Muslim world were to follow them. I believe that the Iranian ayatollahs understand this perfectly well, and in fact are counting on such a strategy.

Psychological breakdown is the only explanation for the lamentable fact that left-liberals, who are no less intelligent than the ayatollahs, cannot see this as clearly as the latter do.

In medical terms, the anti-Zionist stage of Judophobia is a complication that is extremely resistant to treatment.

If we were to attempt to draw a historical analogy, the course of this illness might be compared to the Church's evolving attitude toward the Jews. Like classic liberalism, early Christianity – as taught by Christ – made no distinction between Jews and non-Jews, because in the eyes of God there is "neither Greek nor Jew." However, once Christianity became the state religion of Europe, anti-Semitism became one of the pillars of Church dogma for a full 1500 years. Yet in the second half of the 20th century, at least one branch of Christianity, Catholicism, officially repudiated the idea that Jews bear collective responsibility for the crucifixion of Christ; it condemned anti-Semitism, and apologized for the Church's crimes against the Jewish people.

Such is the history of Catholicism: an initial belief in the incompatibility of anti-Semitism and the new faith; a thousand years of vicious, bloody Judophobia; the beginnings of a gradual recovery.

Can we assume that the course of leftist disease within liberalism will be the same? That from its initial rejection of anti-Semitism, it will pass through an anti-Zionist masquerade, to eventually make a complete recovery?

One would like to hope for an outcome this favorable, but so far there are no signs of improvement. And this is very bad not just for the patient, but for our entire world.

CHAPTER VIII

THEOPHOBIA:
AN INCURABLE
PSYCHOLOGICAL
COMPLICATION OF THE
DISEASE

> *God is not imposed upon us. It is His*
> *distorted, false image that other people thrust*
> *upon us. He Himself, like water, is simply, quietly,*
> *within us. In seeking him, we seek ourselves; in*
> *denying him, we deny ourselves; in mocking him,*
> *we mock ourselves — the choice is ours.*
>
> *Tatyana Tolstaya*

> *Our era is choking itself on hate, because it*
> *has betrayed Christianity.*
>
> *Nikolai Berdyaev*

The militant atheist form of the liberalism's leftist disease is a difficult topic to research — difficult because religious tolerance is always one of classic liberalism's primary principles.

Professor Aleksandr Movsessian formulates this fundamental stance as follows:

Liberalism preaches tolerance of all faiths and points

of view provided that they themselves demonstrate tolerance.

For all its affirmation of tolerance, liberalism nonetheless sees religion as a means of suppressing liberty. In liberal opinion, strict adherence to religious dogmas runs counter to liberalism's chief dogma – individual freedom. Therefore, many liberals share atheist views in the assumption that atheism is least likely to impinge on personal freedom; still others are virulent critics of religion in general.

Let us emphasize that there is nothing wrong with taking a critical view of the postulates of any religion, as long as this critical attitude does not become theophobia at its most painful and aggressive form.

The liberal tradition of skepticism towards Judaic and Christian doctrine dates back to Baruch Spinoza, who laid the foundation for scholarly analysis of Biblical texts. Spinoza was the harbinger of a number of liberal ideas, including that of tolerance. In his *Theologico-Political Treatise*, published in Hamburg in 1670, he spoke out against any sort of religious coercion. The book set off a storm of indignation in religious circles, and Spinoza escaped reprisal only because for some odd reason the treatise was attributed to Hobbes.

This critical approach was, as we have already noted, further developed and refined into finished version of atheism by 18th century French philosophers, who considered religion the chief obstacle on the road to progress.

Yet nonetheless, despite the prevailing current of atheism, classic liberalism never indulged in anti-religious fanaticism. Upholding the concept of freedom of conscience, liberals were tolerant of believers and their "delusions" throughout the 19th century as well.

The idea of engaging in an active war against religion grew not out of classic liberalism, but out of its polar opposites – Marxism and Fascism. Karl Marx and Friedrich Engels saw the dethroning of religion, "the opiate of the people," as one of the chief objectives of proletarian rule. Friedrich Nietzsche cleared the way for Fascist ideology by vilifying the Bible and Christianity in almost gutter terms; he considered both an impediment to the self-affirmation of the Ubermensch, the strong man who, unfettered by morality, relied only on natural instinct.

In the 20th century, the Marxists and the Fascists, who had seized power in Russia and in Germany respectively, put these ideologues' theories into practice.

The Soviet Union, where militant atheism soon became a pillar of state ideology, made particularly short and brutal work of religious institutions. Houses of worship were razed by the thousands; those remaining were often used as warehouses or potato sheds. A few others were converted in more ingenious ways: the government put a swimming pool into the Lutheran Church of St. Peter and Paul in Leningrad; the Naval Cathedral in Kronstadt was gutted and turned into a movie theater and officers club. The pitiful shell of the Russian Orthodox Church preserved for public view was administered and controlled top to bottom by government security services.

An organization called "Union of Militant Atheists" was formed, which published a journal called *The Anti-Religionist,* as well as crude and insulting tracts like Yemelyan Yaroslavsky's "*The Bible for Believers and Non-Believers.*" The promotion of atheism was officially recognized as a government priority in the "Communist education of the working class." Much of this promotion came in the form of terror and repression of those who were reluctant to be "educated." Tens of thousands of clerics were executed or sent to labor camps. Owning a Bible, let alone reading one, was considered counterrevolutionary activity; attending church or synagogue was grounds for dismissal from work or the university. Religious faith was an object of mockery and derision.

These are the realia of militant atheism.

"But what does this have to do with liberalism?" the reader has every right to ask.

With liberalism, nothing. But it has everything to do with the leftist disorder from which liberalism suffers. For there is no evil on this earth that this disease has managed to avoid; there is no infection that the disease has not tried to introduce into the body, even if it runs counter to basic liberal principles. Liberalism has no immunity to the leftist disease, much less to any complications caused by it.

This is precisely the case with the infection called theophobia. The Soviet regime, a breeding ground for atheist obscurantism, the global epicenter of attempts to overthrow Biblical morality, has left the historical arena, taking its militant godless with it. However, the bacilli of aggressive atheism have not disappeared; they have merely found a new host in Western liberalism. In the same way that anti-Semitism penetrated the healthy body liberal by means of anti-Zionism, militant atheism has found its way in also, through the concept of the separation of church and state, and has already begun its work of destruction, the results of which promise to be monstrous.

With the advent of the 21st century, theophobia has begun to spread throughout the United States, a nation with the perhaps most deeply rooted religious traditions in the Christian world.

It started out trivially enough. A certain ascetic-looking gentleman from California demanded that his daughter not be forced to say the word "god" in her kindergarten, and that all references to the Bible be struck from the school's curriculum. After all, since the Constitution of the United States promulgates the separation of church and state, the trappings of religion have no place in public schools. I myself am an atheist, this gentleman declared proudly, and I don't want my daughter to be indoctrinated in religion. The media, eager for a sensation, transformed a run-of-the-mill atheist into a great guardian of the Constitution. Flattered by all the attention, our guardian turned the incident with his daughter into a fundamental question of "the ascendancy of religion in American public institutions." Granted, it later turned out that his wife and daughter weren't living with him, and had no desire to become atheists, but by that time, such petty family problems were barely visible from the lofty heights he had since attained. In 2000, our anti-religious activist filed suit in California, charging that Congress, the President, and the State of California had violated his daughter's rights by including the word "god" in the Pledge of Allegiance. He lost his case in the state courts, then won an appeal. In 2004, the Supreme Court, however, overturned the ruling of the appeals court.

Nonetheless, the wave of anti-religious feeling set off by the California atheist continues to rise. The theophobia he promotes was immediately taken up by left-liberal circles that have no immunity to the disease. Many well known, perfectly respectable liberals have spoken out in his defense – yes, yes, we must preserve the sanctity of the separation of church and state. How outrageous that the President finishes every speech with the words "*God bless America!*" And what about our coins and bills? Why do they say "*In God We Trust*"? What if we don't trust in God, or even believe in one?

As someone who lived through the dictatorship of militant atheism in the USSR, I felt an ominous shiver: "Gentlemen, think about what you're doing, don't cut off the branch that holds you up... This has happened before... and it ended with the most cultured nations of Europe turning into a tame herd of idol-worshippers... I know this all too well... come to your senses... "

But, I repeat, the leftist disease is truly a disease, and there is no cure in sight.

Local activists in Alabama demanded that a privately-funded granite tablet incised with the Ten Commandments be removed from the foyer of the Alabama Supreme Court; this was, they said, a violation of the Constitution. Certain more tolerant parties tried to explain to the activists that the Ten Commandments have long since ceased to be a strictly religious symbol, that they not only underlie universal contemporary moral principles, but are also a historical artifact and a classic work of literature. What is so terrible in reminding people of the abc's of civilized ethics? But no – the activists' attorneys filed suit, and won their case. In November 2003, as if it were something shameful, the granite monument was carted off to a locked room, out of the public eye, and later, in July 2004, was sold to a private party. Roy Moore, the chief justice of the Alabama Supreme Court was removed from office for trying to save it – a clear signal to everyone else who might try to defend the presence of religious symbols in public places.

Oh, how this resembles Soviet methods in the USSR's war against religion!

How did a harmless chunk of granite bearing the Ten Commandments become an offense against the law? What is actually written down in the Constitution about the separation of Church and State? Where is the strict wording that supports this separation?

In the original, 1787 version of the Constitution (without the Bill of Rights), religion is mentioned only once. Article IV states that religious beliefs cannot serve as the basis for appointment for any public office whether state or federal, nor can they be used to deny appointment to a candidate.

Thus, in the Constitution itself, there is not a single word about the separation of Church and State.

In 1791, at the urging of James Madison, the US Congress adopted the First Amendment to the Constitution, which declares that "Congress shall make no law respecting an establishment of religion, or prohibiting the free exercise thereof…"

Thus neither in the Constitution nor in any of the amendments to it is there a single word on the separation of Church and State. Interestingly enough, while the Constitution forbids the establishment of a national church, it does not ban individual states from establishing their own. Religious freedoms protected by the Constitution pertain not to state but to federal law.

Moreover, according to the Constitution, the federal government has no right to forbid individual states to establish their own official religion.

So where did "the separation of Church and State" supposedly laid out in the Constitution actually come from?

In 1802, President Thomas Jefferson wrote a letter to the Danbury Baptist Association of Danbury, Connecticut; in it, he called the First Amendment "a wall of separation between Church and State." Taken in context, Jefferson's phrase movingly underscores the unacceptability of government intervention in matters of religious faith or in the exercise of that faith. However, over time, Jefferson's phrase has led people to believe that this principle is explicitly stated in the Constitution.

Nevertheless, on the basis of a Jeffersonian phrase more emotional than juridical, the leftist malady seeks to instill an intolerance of religion in American society, to build a genuine wall between religion and the state, between religion and the collective society.

I myself have witnessed the recent and rapid spread of the illness. As late as the mid-1990s, religious imagery and symbol was widely displayed throughout Christmas and Hanukkah; it never occurred to anyone that this was somehow unconstitutional or insulting to anyone. I must admit that I liked it: it was a heartfelt and respectful observance of the religious and historical holidays of many peoples.

Over Christmas of 2004, I noticed that these images and symbols were nowhere to be seen in our small Long Island town. I drove through a few other places, and saw that the usual Biblical symbols had been replaced by decorative wreaths or animal figures straight out of some pagan encampment. Genuine salutations like "Merry Christmas" and "Happy Hanukkah" are being replaced by the bland and meaningless "Seasons Greetings."

In December 2005 the ban on symbols of Christmas reached an almost hysterical pitch: no Christmas decorations in school busses, Christmas trees are removed from public places, Santa

Claus and his elves are fired in droves, public school principals and even business owners try to avoid any hint at the religio-historical background of this national holiday. People are afraid to even profess their beliefs in public – God forbid that some activist might accuse them of violating the Constitution.

Once upon a time in the Soviet Union, the Christmas tree was replaced by the New Year's tree; the militant atheists, unable to come up with anything original, simply gave old symbols a new name. Today, American liberals are doing exactly the same, ignorant of the fact that they are simply aping what the Soviets did decades ago.

Anything having to do with the Bible is now under fire; this is the current battlefront.

The fabric of life in a Judaeo-Christian civilization, whatever one's attitude toward religion, is woven of Biblical cloth. Our ethics and our morals, our core beliefs about peace and justice, about personal and social responsibility date back to Biblical texts. Consciously or not, we dot our speech with Biblical imagery; it informs our aesthetics. To excise the Bible from our individual and social life is to change this world so as to make it unrecognizable. One might well take every painting or sculpture on a Biblical theme (or even suggested by one) out of the Hermitage, the Louvre, the Prado, the Vatican, the Metropolitan, the Uffizi and every other museum in the world.

Think about it, ladies and gentlemen. What will we have left?

I venture to suggest that it will be a rather strange and pitiful assortment made up of, say, Kasimir Malevich's black squares, Francis Bacon's foul, flayed carcasses, and Chris Ofili's elephant droppings. Still, I can't guarantee the last; after all, Ofili used the elephant dung to "decorate" an image of the Virgin Mary.

Think again – what will be left, in a world without the Bible? Nothing but elephant dung, perhaps?

Why do left-liberals so stubbornly insist on ridding the society of any mention of the Bible? The answer is very simple. The leftist disease aims to change the very principles on which Western civilization is founded, and like Nietzscheism, sees Biblical morality as the chief obstacle to that change. Nietzsche saw the Bible as an obstacle in the way of the Ubermensch's march to power; left-liberalism sees Biblical ethics as an obstacle in the path to maximal personal freedom, which includes a liberation from the "shackles" of Biblical mores. These false and morbid beliefs in the unrestricted freedom of the individual lie at the heart of these attempts to overturn Biblical mores.

There are many examples of morbid theophobia both in Europe and on the American continent. It may be that the symptoms of the leftist disease are more noticeable in the New World, because the New World, its laws, and its way of life were created by deeply religious people. Among the European immigrants who over the course of three centuries created America, there were many believers seeking refuge from persecution in the Old World. Neither they nor their descendents have any desire to be manipulated by "separation of Church and State."

And nonetheless, this illness within liberalism is advancing in this direction, and advancing quite rapidly...

Friends, instead of "the moral of a story," please allow me to tell you a 20th century fable.

In 1968, American astronauts Frank Borman, Bill Anders and Jim Lovell orbited the moon in Apollo-8.

The US suddenly jumped to the lead in the race to the moon. This was a hard-won achievement for American scientists and engineers; the Soviet Union had always held a significant edge in both rocket science and aeronautics.

The USSR had used its exploits in space to tout the "superiority" of socialism and of Communist morals. "Whoever rules space, rules the world" was the new and improved version of the concept of worldwide Communist revolution.

And of course, as they had ridden the wave of their grand achievements in space, Communist ideologues lost no opportunity to use them to demean religion. The essence of the "space campaign" was best expressed in German Titov's sarcastic remark that while he had spent 25 hours in space, and made 17 orbits of Earth, and had watched very carefully, he never once saw God, or even a single angel.

The West was shocked, less by what the #2 Soviet cosmonaut had said than by the feeling that there was no way to respond to it. He, the atheist, had actually been in outer space, but they, the believers, had not. Seven years passed before believers gained the moral right to offer a worthy and adequate response.

It was simple, but impressive. On December 24, Christmas Eve, 1968, on their ninth orbit of the moon, the crew of Apollo 8 read the immortal first words of the Bible aloud, and radioed them to Earth:

"In the beginning God created the heaven and the earth. And the earth was without form, and void, and darkness was upon the face of the deep. And the spirit of God moved upon the waters."

This was truly a great and shining moment for all of humanity: words written by ancient prophets at the very dawn of our civilization, transmitted from lunar orbit to listeners on Earth. Millions of people at their television, believers and non-believers alike, sensed that something great and mysterious was transpiring before their very eyes, that the wheel of History so stubbornly, inexorably rolling down into Hell had abruptly and unexpectedly turned toward the Heavens.

After taking his turn at reading from the Bible, Frank Borman looked out the window; the lunar night was approaching, and Apollo 8 was about to plunge into darkness again. He heaved a deep sigh, and said:

"And from the crew of Apollo 8, we close with good night, good luck, a Merry Christmas, and God bless you all – all of you on the good Earth."

Hundreds of millions of believers and non-believers on the good Earth held their breath, and felt a lump in their throat...

And now we interrupt this wonderful tale from the past (which readers can find in more detail in R. Zimmerman's book "*Genesis*") to try to answer one question. Could astronauts send such a message to Earth if the story were taking place now, in the 21st century, rather than a mere thirty years or so ago?

This question is purely rhetorical – because we all know without it even being asked that they could not, and if they tried, there would surely be some leftist attorneys ready to file suit against the poor astronauts, as well as the President, the Congress, and NASA to boot!

I can just imagine our suffering patient groaning that "this is outrageous, who gave them permission to read from the Bible on a government spacecraft, over a government channel?! Do we have a Constitution or don't we?! NASA is a government agency! Who gave them permission to send everyone Christmas greetings at the expense of the taxpayer, who might be a non-Christian at all, or even an atheist?"

No, my liberal friends, if American astronauts ever reach Mars, they won't be radioing us the inspired and poetic words of the Bible from that distant planet, because by then we will have a total "separation of Church and State" on what was once the good Earth. At the very best, they'll send us "seasons greetings." As for myself, I probably won't live to see humans fly that far, and thus happily will not have to hear this politically correct vileness transmitted from Martian orbit.

Something wicked this way comes. It is theophobia, an incurable mental disease.

And there is no defense against this disease, and no salvation from it!

CHAPTER IX

TREATMENT OPTIONS?

> *The wise man will seek to avoid infection before seeking to treat it.*
>
> *Thomas More*

> *When a mental illness takes on a violent form that presents a danger to others, the first thing to do is restrain the patient, and only then to think about a cure.*
>
> *Georgy Fedotov*

I would rather not end this brief work on such a pessimistic note, but, to be honest, I cannot offer the reader a politically correct "happy ending."

Let's try to be not merely politically correct, but also to be honest, at least when we're alone with ourselves.

Can liberalism actually be cured of its leftist disorder?

Can it be restored to its original, noble, and healthy state, free of the cancerous metastases of the leftist disease?

We might imagine an individual cure for one or more of the above listed forms of the disease, but it is hard to imagine an effective means of prevention or treatment that would encompass all the complications involved. There are many more than those few described here. Moreover, we know that treating the symptoms rather than the cause of a disease is often unsuccessful. An additional difficulty is that the leftist disorder is a mental pathology, and standard approaches treatment, whether they involve prevention, general counseling, behavior modification or even psychoanalysis, are effective only when applied to those who <u>want</u> to be treated. Unfortunately, this disease is peculiar in that those suffering from it not only do not seek treatment, but also refuse to even discuss the state of their health. They sometimes do not even understand that they are not well, and, as is typical of patients in many psychiatric hospitals for the chronically ill, they think that it is the <u>doctors</u> who are suffering from a mental disease.

So what do we do?

We know how difficult it is to treat diseases once they have advanced to a critical stage, and have presented with a wide range of symptoms. These types of diseases – septicemia, for example – are systemic in nature, and generally have a lethal outcome.

But the systemic nature of the leftist disease manifests itself not just in its multiplicity of symptoms, but also in the extraordinarily great influence of the affected element, be that an individual or a group, on the larger systems of which that element is a part.

If we consider liberalism one conceptual element within a system called Western Civilization, the leftist disease might be defined as a disequilibrium between the internal and the systemic functions performed by those elements; the former are inordinately bloated, which may increase their importance, but which works to the detriment of those functions or tasks that maintain the integrity of the system as a whole. The leftist disorder is undermining the integrity of the social system of the West, and in the final analysis, may lead to the loss of all system function, and thus to the death of the system – something the forces of evil passionately wish for.

How can we restore the balance? How can we diagnose and prevent this left curvature of the psyche before it cripples us entirely?

If anyone knows the answers to these questions, more power to him!

I do not know the answers to these questions, and therefore am generally pessimistic...

Humankind has stepped into the twenty-first century. There are many clouds on the horizon; our children, our grandchildren and our great-grandchildren will be faced with living in a very complex and unforgiving world.

This begs a comparison with the twentieth century, which humankind entered under incomparably more favorable circumstances.

Here is a brief chronicle of the first year of the twentieth century – the year 1901:

> *One major armed conflict: the Boer War in South Africa. Two major terrorist acts in a one year: Russian Minister of Education Nikolai Bogolepov is assassinated by a disgruntled university student; William McKinley, President of the United States, is shot and killed by an anarchist. Queen Victoria dies; her death marks the end of Great Britain's imperial era. Striking achievements are made in the arts: Paul Gauguin, Edvard Munch, Pablo Picasso. August Strindberg writes "Dance of Death;" Anton Chekhov writes "Three Sisters." Ragtime conquers America; Louis Armstrong, future king of jazz, is born. So is Walt Disney. Humankind enters the age of technology: Detroit introduces the mass production of passenger automobiles; Boston does the same for safety razors; England produces the first motorcycle; Max Planck develops radiation theory; Guglielmo Marconi makes the first trans-Atlantic radio transmission; Wilhelm Roentgen wins the Nobel prize. Instant coffee is invented.*

To those who lived at the turn of the century, the future looked bright. Soon, very soon, all these wars would end. Soon peace would reign. Just a little longer, and the limitless powers of human reason would resolve all the problems of the universe; the unprecedented achievements of technology and education would eradicate poverty and injustice, and finally man would no longer be dependent on the forces of nature; he could conquer them, and control them.

This happy-idiot dream was soon to be crushed by the realities of the twentieth century, and most of the dreamers died in the wreckage.

Those who live at the turn of another century have no illusions about a bright future.

Here is a chronicle of the year 2001:

American and British military planes bomb Iraqi military bases. Timothy McVeigh, found guilty of bombing the Murrah Federal Building in Oklahoma City in 1995, is executed. Terrorists attack an international airport in Sri Lanka. Palestinian terrorists organize multiple suicide bombings in towns in Israel, with hundreds of casualties. Kashmiri terrorists attack India's parliament building in New Delhi. Muslim fanatics belonging to a group called Al-Qaeda seize four passenger airplanes; two fly into the twin towers of the World Trade Center; one flies into the Pentagon; the third crashes in a field in Pennsylvania. Three thousand people die. British national Richard Reid tries to blow up a Paris-Miami flight by stuffing explosive into his sneakers. Unknown terrorists send envelopes full of anthrax powder to various US legislators and journalists. Ukrainian military forces, while on a training exercise, mistakenly shoot down a passenger plane en route from Tel-Aviv to Novosibirsk. American military forces invade Afghanistan to overthrow the Taliban regime and destroy the infrastructure of Al-Qaeda; carpet bombing wipes out command points, bases, hideouts, and some of the families of the terrorists. NASA's unmanned spacecraft Galileo passes within 112 miles of Jupiter's moon Io. Microsoft releases another new operating system, Windows- XP. In the US, the first artificial heart transplant takes place.

It seems that the events of 2001 have sobered even the giddiest of optimists.

Of course, it's quite possible that many years from now our children will celebrate the memory of the geniuses born in the first year of the twenty-first century, but even so, it's unlikely that their assessment of the first years of the third millennium will be particularly optimistic.

It is the pessimists and the pragmatists who have set the tone for twenty-first-century generation. And this is due not merely to the sickening threat of terrorism, but to many other things as well: the terrible poverty of underdeveloped nations coupled with uncontrolled birth rates; social conflicts in developed nations; international conflicts; and the total demoralization of both developed and underdeveloped nations.

However, there is another, little-studied reason for the overwhelming pessimism of our age. It is the leftist disease, which plunges even the healthiest and most cheerful of people into deep depression. Watching the news coverage of pacifist demonstrations, listening to statements made by leftist politicians, or seeing the latest work by artists and performers infected with the disease is enough to ruin anyone's mood; it's like peering through the window of a locked psychiatric ward.

Instances of violent psychotic behavior caused by the leftist disease have been recorded. Those suffering from the disease often aggressively chant populist slogans, scream until they lose their voices, and frantically wave their arms.

I fear that this stage of the leftist disease is beyond curing.

"When a mental illness takes on a violent form that presents a danger to others, the first thing to do is restrain the patient, and only then think about a cure. But we must not forget that half the world has come down with this disease, and that the other half is defenseless against infection".

These words were spoken not by a doctor, but by a sociologist and philosopher of religion. Georgy Fedotov was a keen analyst of human behavior, and in this passage he was describing the particularly violent, socialist-fascist forms of insanity that swept over Europe in the mid 20th century.

Oh Lord, how apt these words are in our times, when one half of the civilized world has already caught the noxious leftist disease, and the other half seems defenseless against it.

And this is happening notwithstanding the fact that not only the Islamo-Fascist movement, but other dictatorial regimes intend to exploit this illness for their own criminal ends.

There is a battle raging in the world today, a battle between Islamo-Fascism and human civilization...

Yes, I hear, I hear the skeptical comments of those who prefer to live in ignorance: "No, you're exaggerating, there's no such war going on."

But facts are facts. On October 31, 2005, the *Associated Press* broadcasts the following from Teheran:

> About 300 men and women turned up Sunday at the offices of the Headquarters for Commemorating Martyrs of the Global Islamic Movement to volunteer for suicide bomb attacks against Israel.

"Aaaaha!" the patient exclaims happily. "But that's Israel, what does that have to do with us?" Let's read further, and see what it has to do with us:

> A spokesman for the group said that it had signed up more 45,000 volunteers to undergo training for suicide attacks (45,000 live bombs! Y.O.) since it began recruiting in June 2004... "More than 1000 of them have already been trained. Many of them don't need training since they

are already members of the elite Revolutionary Guards and paramilitary Basij forces", Mohammed Ali Samadi said.

Note that this is not going on in some hidden Al-Qaeda base in the mountain caves of Afghanistan, but in the very capital of Iran (a member of the United Nations), with the moral sanction and monetary support of the Iranian government. And what's curious about all this that many who have joined the war on terrorism pay it lip service, pretend to be fighting all-out against terrorism, hunting down its leaders – but simply cannot find these mythic figures hiding in the hills of Afghanistan and Pakistan. They refuse to see that the epicenter of world terrorism and obscurantism lies right under their noses, in Teheran.

Therefore, let me repeat: a great battle between Islamo-Fascism and human civilization is unfolding. This is not a battle over territory or privilege. It is a battle over the very foundations of our being.

This battle rages not only on the street, when a fanatic sets off a suicide bomb in the midst of a crowd; it is being waged everywhere – in day-care centers and public and parochial schools; in mosques and churches; at universities; at international meetings; at the UN; in the press, on radio, television, and Internet.

The outcome is not so certain as some might think.

The uncertainty lies not in the strength of the forces of evil, but in the inability of good to mount its own all-powerful force against it.

And this inability is in its turn the effect of the senile disorder from which one of the pillars of Western civilization – liberalism – now suffers; that is modern liberalism, which has entered into a critical, chronic and perhaps incurable stage of the leftist disease.

REFERENCES

In English:

James Burnham, «Suicide of the West: An Essay on the Meaning and Destiny of Liberalism», Regnery Publishing, USA, 1985.

Mona Charen, «Useful Idiots: How liberals got it wrong in the Cold war and still blame America first», Regnery Publishing, USA, 2003.

Ann Coulter, «Godless: The Church of Liberalism», Crown Forum, USA, 2006.

Mahatma Gandhi, «The Essential Gandhi», An Anthology, Edited by Louis Fischer, Vintage Books, New York, USA, 1962.

Andre Gide, Retour de L'U.R.S.S., France, 1936.

Sean Hannity, «Let Freedom Ring (Winning the War of Liberty over Liberalism)», Regan Books, USA, 2002.

Eric Hoffer, «Israel's Peculiar Position», Los Angeles Times, May 26, 1968.

Nils Karlson, «European Socialism», The Ratio Institute, The Architecture of Modern Political Power (**AMPP**), April 28, 2004.

Yuri Okunev, «The Lost War», AuthorHouse, USA, 2004.

Yuri Okunev, «Left-Wing Liberalism: A Senile Disorder», Xlibris Corporation, Philadelphia, USA, 2007.

Yuri Okunev, «The Axis of World History», Xlibris

Corporation, Philadelphia, USA, 2008.

George Orwell, «Pacifism and the War», London, UK, 1942.

Ion Mihai Pacepa, «Red Horizons», Regnery Gateway, Washington, USA, 1990.

Richard Pipes, «Dissecting Modern Liberalism», National Review, USA, December 19, 2005.

Sebastián Vivar Rodríguez, «Europa murió en Auschwitz», 21/11/2004, http://www.gentiuno.com/articulo. asp?articulo=1865

Michael Savage, «Liberalism is a Mental Disorder: Savage Solutions», Thomas Nelson, Inc., 2005.

Neville Williams, «Chronology of the 20th Century», Helicón Publishing Ltd., UK, 1995.

Alexander Yabrov, «How Man Exists», 1-st Books Library, USA, 2001.

Robert Zimmerman, «Genesis», Four Walls Eight Windows, New York-London, 1998.

In Russian:

Бердяев Н.А., «Избранные произведения», Феникс, Ростов-на-Дону, РФ, 1997.

Большой энциклопедический словарь, Москва – С.-Петербург, РФ, 1999.

Всемирная энциклопедия: философия, АСТ–Москва, ХАРВЕСТ–Минск, 2001.

Гордин Я.А., «Лев Толстой и русская история», Hermitage Publishers, USA, 1992.

Марк Зальцберг, «Goodbye, America – США начинают строить коммунизм», Новая Газета, Интернет-издание:

www.novayagazeta.ru

Жан-Франсуа Лиотар, «Хайдеггер и евреи», Аксиома, Санкт-Петербург, РФ, 2001.

Мовсесян А.Г., «Либерализм и экономика», Москва, «Логос», РФ, 2003.

Андрей Московит (Ефимов И.М.), «Метаполитика», Strathcona Publishing, USA, 1978.

Фридрих Ницше, «Воля к власти», АСТ–Москва, ХАРВЕСТ–Минск, 2000.

Окунев Ю.Б. «Старческая болезнь левизны в либерализме», Поверенный, Рязань, РФ, 2004.

Окунев Ю.Б., «Ось всемирной истории», Искусство России, Санкт-Петербург, РФ, 2004.

Окунев Ю.Б., «Ось всемирной истории», Издание второе под редакцией И.М. Ефимова, Hermitage Publishers, USA, 2006.

Советский энциклопедический словарь, Издание второе, Москва, СССР, 1983.

Толстой Л.Н., Антология мысли, Эксмо-Пресс, Москва, РФ, 1998.

Федотов Г.П., «О святости, интеллигенции и большевизме», Санкт-Петербургский государственный университет, Санкт-Петербург, РФ, 1994.

Лион Фейхтвангер, «Москва 1937», Перевод с немецкого, ГИХЛ, Москва, СССР, 1937.

Георгий Чернявский, «Высокоинтеллектуальные слуги Сталина», Газета Каскад, №190, май 2003, Балтимор, США, www.kackad.com

Шаламов В.Т., «Колымские рассказы», Советская Россия, Москва, РФ, 1993.

Эренбург И.Г., «Люди, годы, жизнь», Советский писатель, Москва, СССР, 1990.

ABOUT THE AUTHOR

Dr. Yuri Okunev is a scientist in the field of communication technology – the most modern area of applied mathematics and physics. He is regarded as one of the pioneers of the Information Century, and his scientific school found a wide application throughout the world, including such countries as Russia, USA, and Israel.

After graduation from the St.Petersburg State University of Telecommunications, Russia, Okunev organized a scientific laboratory, which soon became a leading research center of a worldwide scale. Among many important directions, the laboratory developed new methods of long-distance digital communications and satellite communications. The laboratory pioneered in wireless technology and received a worldwide recognition for innovations in the field of phase-difference modulation techniques. In Russia, Okunev published 14 scientific monographs and manuals.

In 1993 Yuri Okunev moved to the United States and since then has been working in the American telecommunications industry. Working for leading research centers, such as Bell Labs of Lucent Technologies, PCTel, Symbol Technologies/Motorola, he participated in the development of advanced wireless systems and technologies. In the USA, Okunev has published 25 patents and a new scientific monograph *Phase and Phase-Difference Modulation in Digital Communications*.

In 2007 Dr. Okunev received a prestigious IEEE Charles Hirsch Award for «outstanding contribution to the phase modulation theory and wireless system design».

Dr. Okunev's creativity, however, reaches beyond his professional accomplishments. An outstanding philosopher-humanist, Okunev has authored deeply thoughtful books and essays of socio-philosophical nature. As anyone who has read his books can attest, Okunev is a profound thinker and a masterful writer. His books *Letters to Loved Ones from the XX Century*, *The Axis of World History*, *Left-Wing Liberalism: A Senile Disorder*, and *The Lost War* are important contributions to treasury of the world socio-philosophical literature.

The Axis of World History by Yuri Okunev was recognized by USA Book News as "Award-Winning Finalist in the World History category of the National Best Books 2008 Awards"